The Environments of Architecture

This well-illustrated work provides a much needed and topical philosophical introduction to the place of environmental design in architecture.

The Environments of Architecture sets out a range of considerations necessary to produce appropriate internal environments in the context of a wider discussion on the effect of building decisions on the broader environment.

The authors, from architecture and engineering, both academia and practice, provide a rounded and well-balanced introduction to this important topic. Starting from a belief that the built environment can contribute more positively to the planet and the pleasure of places as well as answering the practical demands of comfort, they cover site planning, form, materials, construction and operation as well as looking at design on a city level. Challenging the mechanical model of architecture that prevailed in the twentieth century, they explore analogies with biological models that might be more appropriate for the twenty-first.

Presenting a thoughtful and stimulating approach to the built environment, this book forms an excellent guide for practitioners, students and academics concerned with our built environment.

Randall Thomas is a consultant to Max Fordham LLP and teaches in the UK and France. He is currently a Visiting Professor at the University of Cambridge, Professor of Sustainable Environmental Design at Kingston University and a Course Organiser at the AA. His previous books include *Environmental Design*, *Photovoltaics and Architecture* and *Sustainable Urgan Design*. **Trevor Garnham** is an architect and Principal Lecturer at Kingston University School of Architecture. His previous books include *Lines on the Landscape*, *Circles from the Sky*; *Monuments of Neolithic Orkney* and three books on Arts and Crafts buildings. He is a regular reviewer of contemporary buildings for architectural journals.

The Environments of Architecture

Environmental design in context

Randall Thomas (Consultant, Max Fordham LLP) and
Trevor Garnham (Kingston University, UK)

LONDON AND NEW YORK

First published 2007 by Taylor & Francis
Published 2014 by Routledge
2 Park Square, Milton Park, Abingdon, Oxon OX14 4RN

Simultaneously published in the USA and Canada
by Routledge
711 Third Ave, New York, NY 10017

Routledge is an imprint of the Taylor & Francis Group, an informa business

Typeset in Akzidenz by Wearset Ltd, Boldon, Tyne and Wear

British Library Cataloguing in Publication Data
A catalogue record for this book is available from the British Library

Library of Congress Cataloging in Publication Data
Thomas, Randall.
The environments of architecture : environmental design in context /
Randall Thomas and Trevor Garnham.
p. cm.
Includes bibliographical references and index.
1. Architecture–Environmental aspects. I. Garnham, Trevor. II. Title.
NA2542.35.T48 2007
720'.47–dc22
2007005319

ISBN13: 978-0-415-36088-3 (hbk)
ISBN13: 978-0-415-36089-0 (pbk)
ISBN13: 978-0-203-79940-6 (ebk)

Contents

Illustration credits

ASHRAE Publications 5.9
Mohammed Ageli 2.4
Tony Aldrich 3.17
Richard Bryant 3.18
David Bodenham A3b
Mike Caldwell/Paul Vonberg 6.9
Peter Cook 6.11
Bill Dunster Architects 8.18
Kris Ellam 3.17, 4.10, 4.22
Max Fordham LLP 5.5, A4, A6, A7
Timothy Garnham 2.17, 3.7, 3.11, 3.14, 3.15, 4.7, 4.15, 4.24, 7.14, 8.17
Trevor Garnham 2.8, 2.9, 2.10, 2.16, 2.18, 3.2, 3.4, 3.5, 3.7, 3.13, 3.17, 4.1, 4.2, 4.9, 4.12, 4.14, 4.18, 4.19, 4.24, 6.5, 6.12, 7.6, 7.10, 7.12, 7.16, 7.17, 7.18, 7.19, 7.20, 8.4, 8.7, 8.8, 8.9, 8.14, 8.20, A3c
Bedford-Lemire & Co. Ltd. 7.23
Heritage Services, Bath and North East Somerset Council 6.8
Ian Lawson/Max Fordham LLP 6.13
Lorna Lucas 2.10
Los Angeles County Museum of Natural History, Marion Boyars Publishers 5.11
MacCormac Jamieson Prichard 8.30
Jon Miller, Hedrich Blessing, Landmarks Preservation Council of Illinois 5.12, A3e
The National Gallery 7.4
Nature Publishing Group B1a
Will Perkis 3.16, 7.15
Hareth Pochee/Max Fordham 8.27
Jo Reid and John Peck 6.7
Phil Sayer 8.18
Heini Schneebeli 4.16, 7.13
Shakespeare Birthplace Trust Records Office 5.4
Mark Skelly Cover page
Timothy Soar 5.15
Anna Stanford 5.16
Oak Taylor-Smith/Max Fordham LLP 6.14, 8.31
Paul Smoothy A3a
Anne Thomas 8.26
Randall Thomas 3.1, 7.24, 8.23, 8.29, A3f

Bill Ungless 2.6
Bastian Valkenburg 4.11
Tim Wilcockson 2.15, 7.22
Frank Lloyd Wright 6.10
All line illustrations drawn by Trevor and Timothy Garnham.

Preface

Although we had both been teaching independently at Kingston for some years, it was the appropriately named Contact Theatre that drew us together. From this came a course we taught jointly and which culminates in this book. Trevor's review of the Contact[1] led to a series of lively discussions. We both had firm convictions (if not a messianic belief) that environmental considerations could lead to an architecture and urbanism of new forms and materials.

We were both deeply dissatisfied with how the environment was being treated. We, as all card-carrying utopists, believed that it could be better in the future although we disagreed on whether it was (much) better in the past – and, one might ask, which past? We had a love of the natural world, of its beaches, deserts, mountains, forests, meadows and plants. We believed that the built environment could contribute to the pleasure of places as well as answering the practical demands of comfort.

In spite of our varied academic and professional backgrounds, we shared elements of a common cultural background including the writings of C.P. Snow,[2] Sigfried Giedion, Reyner Banham, Seamus Heaney and many others. Our views were often different, at times distressingly so. This is perhaps unsurprising in encounters between English Romantics and American Pragmatists.

Despite our differences, we entered into a Faustian bargain (it wasn't quite clear who played which role) to test our ideas and write this book. What ideas? A plethora of ideas ranging from the influence of Heidegger, Kant's idealism, the role of Bacon in the development of science, the origins of symmetry, the development of the grid, whether the Seven Dwarfs used as caryatids contributed to architecture, and several hundred more. Clearly, though, a book of only 200 or so pages was going to require a full box of Occam's razors.[3]

So we set out in pursuit of essential points. What really mattered? And what book would we ourselves have wanted to start our introduction to architecture and environmental design with? Rather than another detailed design book – there are many of these and some will be referred to here – we felt that there was a place for a more philosophical introduction which would appeal to students of architecture and all those interested in the environment.

One of our aims is to draw upon our different background knowledge and ways of thinking in the humanities and sciences in the conviction that both are essential for an understanding of architecture and the creation of a sustainable and satisfying environment. Art and science are both 'the tools of life' as Ortega y Gasset well said.

And we agreed that words were important tools. We shunned Humpty Dumpty's view that 'When I use a word it means just what I choose it to mean ...'. Words fix things and they contain the memory of mankind's thinking about the matter of the world. A careful consideration of the meanings carried in words can be a useful aid to clear thinking about a subject.

Presenting a thoughtful, wide-ranging and stimulating approach to the built environment in a very broad context was our goal. Our themes pop up and down throughout the chapters, sometimes winking, sometimes screaming. The result is not a fully integrated, seamless work of complete resolution – architecture, buildings and cities should not strive for that, we believe, nor should books about them. Our collective future is a debatable one and unresolved. This book is for everyone who looks forward to a bit more poetry, architecture and a polar bear or two.[4]

Acknowledgements

First of all, the authors would like to thank Caroline Mallinder of Taylor & Francis for bearing with them during the long period that it took to complete this book to their satisfaction. We would also like to express our appreciation to Georgina Boyle of Wearset Publishing Services for her care in the production of the work.

It is a pleasure to acknowledge the contribution of Joanna van Heyningen, Stephen Pretlove, Nigel Craddock, Bill Ungless, Rod Mulvey and Elie Zahar for their careful reading and invaluable comments on the drafts. Any errors remain the sole responsibility of the authors, but we believe that the assistance of our colleagues and friends will have gone some way towards eliminating most of them.

We would like to thank Timothy Garnham for using his digital and newly-acquired graphic design skills to turn some very basic line drawings into fine illustrations.

For over three years Anna Georgiou has been a calm and encouraging midwife for the text and her administrative and political skills have been invaluable.

Randall Thomas would like to thank Michèle, Anne and Dom for lending their sympathetic ears to an obsessed author and Kingston University for its support. Trevor Garnham would like to thank Lorna for her calmness and understanding.

Finally, it is perhaps most appropriate to acknowledge the contribution of our students at Kingston, who were a both patient and stimulating test bed for our ideas.

Note to readers

One intention of this publication is to provide an overview for those involved, professionally, as students, or in any other way, with buildings. It is not intended to be exhaustive or definitive and it will be necessary for users of the information to exercise their own professional judgement when deciding whether or not to abide by it.

It cannot be guaranteed that any of the material in the book is appropriate to a particular use. Readers are advised to consult all current Building Regulations, Building Standards or other applicable guidelines. Health and Safety codes and so forth, as well as up-to-date information on all materials and products.

Physics units, conversion factors and abbreviations

Units and conversion factors

The unit of thermodynamic temperatures in the SI system is the kelvin (K). For this reason derived units such as thermal conductivity are expressed as watts per metre kelvin (W/mK). However, the Celsius (°C) temperature scale is also in common use (the Celsius scale is also known as the centigrade scale). Absolute temperature in degrees kelvin is found by adding 273 to degrees Celsius. Thus, 30°C + 273 = 300K.

Power

W = watt (1 W = 0.86 kcal/h)
kW = kilowatt
kWe = kilowatt of electrical output

Energy

kWh = kilowatt hour
MWh = megawatt hour (1,000 kWh)
GJ = gigajoule (1 GJ = 278 kWh)

Heat transfer coefficient

1W/m^2 K (1 watt per m^2 of area per degree K of temperature difference)

Abbreviations

dB = decibel
ha = hectare
Hz = hertz (1 Hz = 1 cycle per second)
kcal = kilocalorie
nm = 1 nanometre

At a few places in the text, to refer to one of the authors we have used our initials TG and RT.

Glossary

Absorption chiller units These cooling devices use refrigerants in an absorption cycle (rather than the more common compression cycle found in, for example, household refrigerators). One reason that they are of interest is that they can use solar energy in the part of the cycle in which the refrigerant is regenerated.

Combined heat and power Refers to the combined production of heat and electricity. Equipment which can do this varies from traditional fossil-fuel driven generators which produce electricity and from which heat is recovered to fuel cells (see below).

Coolth A term used loosely to describe the fact that the temperature of a body can be lowered and advantage taken of this to provide cooling. Thus, for example, if cold night air is introduced into a space, it is said that 'coolth' is being stored to help prevent the space from becoming too warm the following day.

Daylight factor Broadly, the daylight factor is the ratio of the light level inside to that outside.

Dry air coolers Heat exchangers which use ambient air to cool a circulating fluid. (The term 'dry' is used to distinguish them from a group of other heat exchangers known as 'wet' cooling towers which unsurprisingly use water and air.)

Fuel cells There are many types of fuel cell but the ones of most interest here combine hydrogen and oxygen to produce electricity (d.c.) and water, giving off heat in the process.

Phase-change materials Many materials undergo a change of phase, for example, water (liquid) can change to ice (solid). Materials that undergo a change of phase in a suitable temperature range for buildings may be useful for energy storage. Glauber's salt (crystalline sulphate of sodium), for example, decomposes at 32°C.

Schlieren photography This technique, whose name is derived from the German word for streaks or striations, uses a system of grids and a lens to record photographically the refraction of light rays by thermal gradients in the air.

1

Introduction

Architecture needs to be located in an environmental, historical and cultural context. Our environmental context is one of rampant energy consumption, dwindling fossil fuels and global warming. Buildings are a major consumer of energy in both their construction and operation. For example, 40–50 per cent of all energy consumed in Europe is used in buildings and 40–60 per cent of this is for heating and ventilating. As such they are a significant cause of our environmental problems. However, it is possible through design (and management) to create an architecture (and here it is understood that we include urban design) which contributes more positively to the planet and our lives. The environmental impact of a building and, in particular, its CO_2 emissions, depends on its overall design – site planning, form, materials, construction and operation all affect performance.

To succeed in developing a new architecture a deeper understanding of science, history and culture than is evident at present will be required. The built environment of the past was at times the creation of geniuses like Brunelleschi and Wren who combined the humanities and sciences. More recently Le Corbusier (along with Ozenfant) expressed an interest in both fields. But very broadly there are significant differences – science is often a more collective effort and much of it depends on a quantitive understanding of the physical world and art is usually more individualistic and qualitative. A scientific theory can normally be objectively verified – an aesthetic theory is more subjective. Science can be seen as a self-correcting thought system which advances but literature, say, has a more active historical sense of the past which shapes the present.[1] Architecture needs to combine the understanding of the relationship between past and present that literature, for example, has with the knowledge that comes from science (and engineering).

Design is about the future, but to be human is to be poised between the past and the future. This very moment and any human activity are made possible by memory and anticipation. Hence we agreed that we needed to cover a sense of time, of history and of the future. Our two-sided Janus approach combines end-of-millennium malaise with a desperate desire to be optimistic. So this book looks both backwards and forwards, embracing both tradition and science. Architecture should learn from how vernacular buildings – the work of anonymous builders – responded to the environment as well as up-to-date methods of design prediction.

It must take a broad approach, almost an anthropological one. Anthropology is 'the science of ... mankind in the widest sense',[2] and can be drawn upon, for example, in viewing the role of buildings – as all material culture – to be as much symbolic as practical. Cultures express how they have adapted to the environmental conditions of a particular place as well as the universal, if extremely varied, belief in the spiritual or poetic. Environmental factors have, of course, been major determinants of form and the range of traditional, vernacular buildings records mankind's adaptability, invention and capacity to produce many kinds of beauty. The sarawak house and the igloo are as different from one another as the parrot and the penguin.

Our view is that it is useful to juxtapose science with well-adapted vernacular buildings. Architecture is all too often seen primarily either as a technical matter or one of personal visual expression and novel form making rather than a cultural concern, as it should be. Sustainable design should aim to make a cultural contribution, which entails combining a scientific understanding of environmental principles with a sense of how vernacular buildings made comfortable living conditions before mechanical services. Kenneth Frampton's concept of 'Critical Regionalism' has been important in establishing an intellectual basis for the approach outlined in the following chapters, dealing with the issues of the age whilst locating building design in a cultural and environmental context. It helps to overcome the dangers of nostalgia and simply reviving the 'forms of a lost vernacular'.[3] Critical Regionalism is a strategy that Frampton proposes for a modern architecture that resists a bland, homogeneous global culture by responding carefully to particular qualities of a region – climate, materials, light and cultural traditions.

One of the most crucial questions for architecture and environmental design is 'What is the relationship among things, organisms and their context?' This complex question requires an equally rich response. The answer is not merely a quantitative one, which, for example, analyses energy flows in a community, but a qualitative one which gives full consideration to poetry, beauty and delight (and in this there will also be decay, sadness and death). Perhaps it is not possible to teach these things and perhaps, strictly speaking, there can be no textbooks on design, for each problem presents itself anew and each generation has new ideas to express. But one can set up some signposts, and that is what we have tried to do below.

A related question to the one above is to what extent biology can be of service to architecture. Our view is that simple biological analogies are of limited value but an analysis of the scientific bases – the physics, the chemistry, the ecology and so forth – can be very rewarding.

Views of science and technology are complex and varied (including ours). Although science clearly has the capability of making our lives better, whether in fact it will is less so. Science is often confused with technology. Science can help us understand the environment and develop solutions to its problems. Technology, which can be seen as the application of science, can have both positive and negative effects on the environment as we know from dams, nuclear power plants, genetically modified crops, plastics and on and on. We need to use the right (sometimes known as 'appropriate') technologies and, as might be suspected, there is not unanimity on what these are. However, we are of the opinion that the age of mechanisation, as described intelligently and eloquently by Giedion in *Mechanisation Takes Command* and Reyner Banham in *The Architecture of the Well-tempered Environment*, is giving way to a more biological approach which values life more and views the built and natural (including the physical) environments as one. This is one basis for a sustainable architecture which can be an element of a new broad understanding of mankind's relationship to the planet.[4]

Much of this discussion is intertwined, of course, with functionalism, which was taken up in a somewhat simplistic way in modern architecture. Evolution produces organisms that function well (or as well as they have to) and so must buildings in relation to their environment. The built environment needs to be designed in the full knowledge of the impact human actions have in a broader context. Building in this way will be well-adapted to its local environment, its 'habitat', and will have internal arrangements that efficiently deliver essential 'nutrients', fresh air, warmth, light, etc. But meaning, per se, is absent in evolution. We humans inherit a world where things have significance for us – we find or add meaning and cannot live without it. Meanings may be as many and varied as there are individuals. But we share a common humanity and all live on planet Earth alongside the animals and plants. Considered in this way a 'deep structure' might be glimpsed that suggests how the built environment might be more meaningful for us and at the same time less detrimental to the rest of life on the planet. An important role of architecture and environmental design is to ensure that we adapt at least most of our human habitat as symbiotically to the broader environment as naturally many plants and animals do. One key to this will be the use of renewable energy sources. If the nineteenth century was the age of coal and the twentieth of oil, the twenty-first will be the age of the sun.

The built environment in which we live has been shaped by two dominant traditions – a vernacular one which tends to be protective of its inhabitants and that of the Modern Movement which aimed at an 'open' architecture of light and space. There is a need to combine the best elements of both these influences and this will be done in part through a careful study of form. Morphology is one of the principal themes in design and will be a key aspect of sustainable architecture.

For us, good architecture embodies our thinking about the world and our place in it. This, we believe, needs to be a place of respect for humans and the environment (both physical and biological). How to achieve this in a rapidly urbanising world is an exceptional and urgent challenge. The worldwide urban population grows by one million each week. Cities now form the environment for most of humanity for most of the time and by mid-century we may be nine billion people with 70 per cent living in cities. The impact of cities on the landscape due to their consumption of energy and materials and their production of wastes is untold. It is impossible to consider the future of architecture without a concept of what urban life may be.

Taking up the themes above, the next three chapters deal with site and setting and building design to develop ways of viewing the site as a specific place on the Earth and at one moment of its history. The first discusses how comfort can be provided (or more broadly how organisms adapt to and alter their environments). The two chapters which follow start to examine how mankind has created buildings for sites in both the past and present and how it may do so in the very different environmental context of global warming and dwindling fossil fuel resources. The three after these go into more detail on specialised areas of environmental design – heating, cooling, ventilation and lighting. Although these are given separate chapter headings, it is worthwhile noting that they are all part of a unity which runs from site to detail and that successful architecture can not divorce form, structure and services. In this again the biological unity of organisms with their integrated form, structure and physiology has inspired us. The penultimate chapter is on cities and reflects upon key points from the preceding ones in relation to groups of buildings in an urban context.

2

Site and setting

Introduction

Architecture is inextricably bound up with place. A building is constructed in a particular place, whatever other factors shape a design. Hence the siting of a building – its relationship to its immediate environment – is of enormous importance. In addition, a sustainable architecture must consider the impact on the broader environment. The word site is derived from the Latin *situs*, which translates as 'local position'. But *situs* is connected to *sinere*, which means 'to leave, or allow to remain'. In this we detect a resonance between site and sustainability, that a building should adapt itself to the inherent features of a place as much as it responds to internal dictates of function to develop form. Interestingly, the word sustainability comes from the Latin *tenere*, which means 'to hold firm', a root which gives us tender, to attend upon, and to be attentive to, all words usefully kept in mind as we consider our intentions in siting a building and developing a sustainable architecture.

Biology and biodiversity are the keys to our sustainable future. This starts at the site level in its broadest sense. Before building, a site is what its geology and history have made it, and it is the habitat of plants, insects, birds and animals.[1] The earth teems with life. There is one basic connection between all of its myriad forms; they use material and energy from their surrounding environment to grow. What is of particular interest to us is that in organisms, form, materials, physiology and, in the case of animals, behaviour constitute a whole that is adapted to the environment.

Plants absorb sunlight and use its energy to photosynthesise the food they need, a process that involves taking up carbon dioxide from the air and drawing up water with dissolved chemicals from the earth. Animals get their energy by eating plants, and so taking over their carbohydrates and other nutrients, or by eating other animals. An animal is a bit like a tube that takes in food and air at one end and expels waste at the other. Buildings are not unlike animals in this respect, taking in energy, light and air and producing waste. But buildings are also like plants in being rooted in the ground and rising to the light and air. Sustainable architecture for the twenty-first century will benefit by drawing analogies with biological systems rather than mechanical ones, which featured large in the twentieth.

Life depends on the energy that the Earth receives from the sun. The Earth's atmosphere allows visible light, infrared and a small amount of ultraviolet to reach the surface. The sun's radiation warms both the world of the living and the mineral. When life began on Earth about 3.5 billion years ago, the atmosphere was very different from today – the likely mixture of methane, hydrogen sulphide and carbon monoxide is toxic to life as we know it now. As early as 2,500 million years ago cyanobacteria were harvesting light to make sugars from carbon dioxide and water, and in this process of photosynthesis producing oxygen, which consequently transformed the atmosphere. Eventually our current atmosphere developed which is broadly suitable for life on Earth (the number of people who die from respiratory diseases is one reason for the qualification). Although carbon dioxide constitutes less than 0.05 per cent of atmospheric gases, almost all of the carbon in living things is derived from this. The atmosphere also makes life possible and sustainable for, along with the Earth's rate of rotation, it prevents our planet becoming impossibly hot or cold.

Just as animals need food for energy – partly to maintain their body temperature – so buildings require energy to keep them warm (or cool in a hot environment), as well as light and air. The basic energy source for buildings, as for animals, is plants – fossil fuels, coal, oil and gas being the product of long dead vegetation. As is well known, the proportion of carbon dioxide in the atmosphere is rising steadily and this is a principal cause of global warming. This is largely a result of burning fossil fuels, burning in just a few decades the remains of plants that absorbed the gas over millions of years. The energy used in heating and cooling buildings is a major factor in this. The reserves of fossil fuel are running down but new technologies are emerging that will allow buildings to become more like plants and produce energy from solar radiation. Also older forms of energy production are being re-deployed: windmills becoming wind turbines, water mills adapted to produce hydropower, etc. Recent advances in technology that allow buildings to generate energy add a new dimension to the more traditional demand whereby the siting and form of a building, its orientation and design make optimum use of the microclimate of a site.

Until quite recently people tended to remain in the locality where they were born.[2] Like plants, a culture and its buildings became rooted in a particular place. Any place is made up of the interaction of many factors: topography, geology, soil chemistry, climate, vegetation, human history, culture, time and chance. Locally available natural material – stone, earth, clay, trees and plants – were traditionally used to make congenial habitats for humans wherever they settled. These traditional or vernacular buildings often strike us as having an organic relationship with the environment in which they sit and often contribute greatly to a sense of place.

In contrast, mobility has become such a feature of modern life that with our inexpensive – if environmentally costly – travel, people experience at first hand a range of cultures and climates (mostly people from cold northern climates seeking the sun and warm seas of the south). Our ability to travel freely was reflected in twentieth-century architecture that became international, shaped by concerns for function, industrial production, and the assumption that buildings could rely on cheap energy and mechanical systems for heating, cooling, ventilation and lighting. The modern, Western way of life and consumer lifestyle is inextricably bound up with the environmental crisis reported daily in our newspapers. A primary aim of a sustainable architecture must be to reduce energy consumed by buildings and hence their negative impact upon the broader environment.

In one sense, as we said in the Introduction, there can be no textbooks on design if for no other reason than, while some design problems recur, every site is unique. But buildings, like plants and animals, can be thought of as similar to particular species with some shared characteristics. Morphology is critical, and runs through this book from site to city. One sees the same phenomena at varying scales, from organism to site and from house to city. And this is because the same laws of physics apply. For makers of buildings, physics is a handmaiden of design, but for the biological world it is the master. A major aim here is to help develop a sensibility in building design that is grounded in a greater sensitivity to the broader environment. Like plants and animals, a building must adapt itself to its habitat. But at the same time it must make an appropriate habitat for humankind. This chapter aims to help us understand how buildings should relate to their context in part by looking at the way plants and animals adapt to their environments. Traditional or vernacular buildings evolved to make comfortable homes for humans all over the Earth and examples of these will be looked at for guidance in conjunction with the deeper understanding of the environment provided by modern science.

One starts with a holistic view and a local understanding of the region, climate, geology, materials, biodiversity, etc. Then we might think of a site as the centre of an imaginary sphere say 100 m in diameter and consider the factors impinging upon the proposed building at the same time as assessing the building's impact upon the existing place. We must consider how solar potential is maximised, how to make use of rainwater, draw upon the potential of the ground – not only its topography but also its groundwater and thermal capacity – and work with the wind. It is easy to give insufficient attention to these factors when we design on our sheets of white paper or computer screens in the knowledge that with the flick of a switch we can have light, air and the desired temperature. Instead of such detached design processes, we need to imagine, for example, the path of the sun, using data and drawings, during any one day from dawn until dusk, and also its changing altitude from winter to summer (Fig. 2.1). Vernacular buildings were passive modifiers of the environment whereas new technologies – in particular photovoltaic cells that convert solar energy into electricity – invite buildings to be active participants. Solar energy is set to play an increasing role in generating the form, and affecting the appearance and construction of buildings. Vernacular buildings conserved energy, new buildings can generate energy. A sustainable architecture looks at the potential assets of the site and also relies on sustainable energy sources (as was the case in the past – at least to a certain extent) and this will affect site planning.

Our response to places is often an emotional or intuitive one. We might love the sea or we might prefer hills to mountains. We might like foreign cities or prefer the countryside, enjoying the food and admiring buildings of a region. As people we see the poetry in a place, but as designers we must also consider the practical aspects of site planning. As designers of the environment we have to be attentive to both the qualitative and the quantitative aspects of place making. Any human situation entails considering the poetic and the practical simultaneously, creating the appropriate atmosphere as well as the physical conditions for material comfort.

In looking at how to design a building that fits its site, we might remember that architecture is fundamentally about space and relationships. A design must draw what is needed from the vicinity and at the same time take account of the impact the new structure will have on the quality of its immediate surroundings. People inhabit their buildings and other forms of life find their habitats remade by them. Good design should aim to help people relate to the particular qualities of the world as well as being practically sustainable.

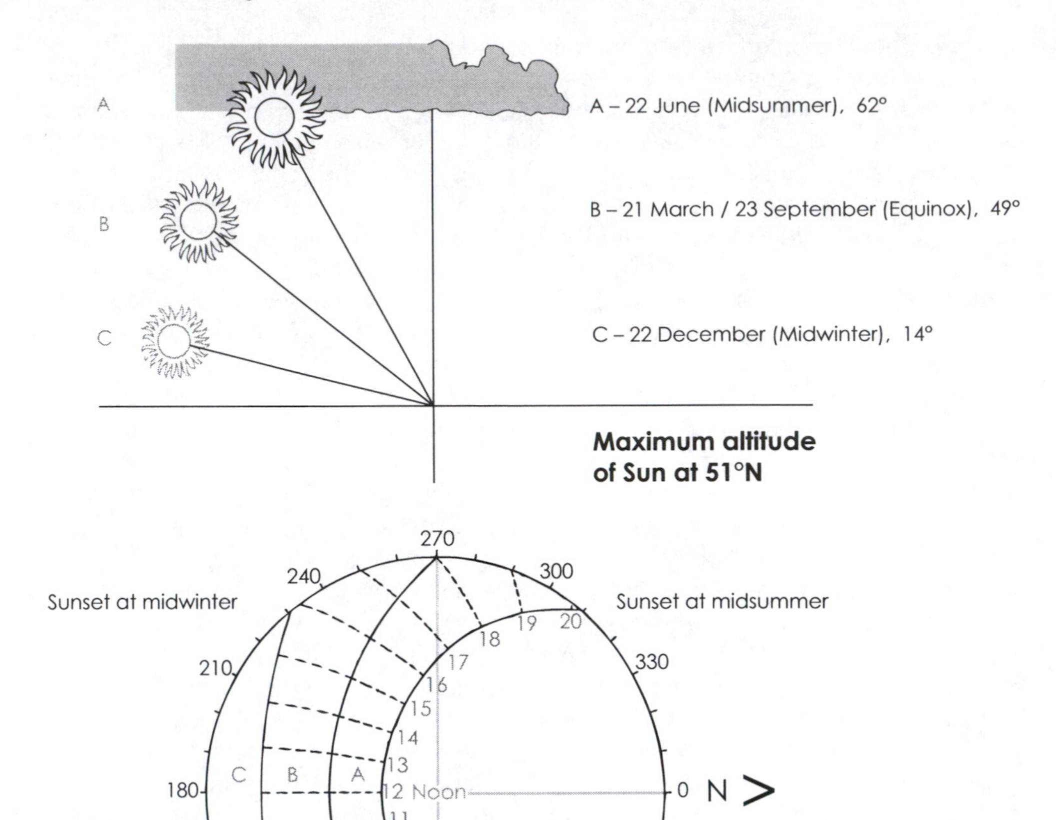

2.1 Sun path diagram – based on Pleijel's stereographic sun path diagram for 51°N – e.g. Greenwich. Within the circle is shown the sun's path from east to west, sunrise to sunset, at three significant times of the year. Immediately above is shown the altitude of the sun at noon at these times of year.

The human need for shelter

All living creatures have adapted to their environments. Polar bears and seals, for example, have fur and a thick layer of fat to survive in the extremely cold Arctic climate. Camels have their humps to store water, which helps them survive in the hot, dry desert. The first recognisable humans – *Homo habilis* and their descendants *Homo erectus* – evolved in the savannah region of Africa where the temperature and vegetation suited their biological needs. Unlike many animals that adapted to a specific environment and remained there, humans eventually spread all over the Earth and had to adapt their lives to every climatic region. In the earlier days of humankind people chose the best location for building with this in mind. Nowadays when humans cover the Earth, sites for building are usually presented to us, so we have to work with the constraints and possibilities of any given particular location.

Biologists use the term *biomes* to designate the different regions of the Earth with their characteristic climates, day-length, topography, vegetation and animals. The UK is a region of 'Temperate deciduous forest', sometimes sub-divided into 'Mid-European coastal' for England and Wales, and 'North-European coastal' for Scotland.[3] Climate is determined by several influences including: distance from the Equator, direction of wind, distribution of land mass, the proximity of water and mountains. In its simplest form, we can picture the *biomes* as roughly parallel zones, like lines of latitude, running from the Equator to the Poles: tropical forest, savannah (dry grassland with trees), hot desert, temperate grassland, temperate forest and cold desert or tundra as we near the Arctic.

Over millions of years, plants have adapted to an extreme range of climatic conditions, often in the most extraordinarily inventive ways. In tropical rainforests, for example, trees endure frequent torrential rain and some have evolved small leaves with a distinct point at the tip to throw off water quickly, rather like a gargoyle on a medieval building. Temperate forests cover wide areas north to south and include broad-leaved trees and conifers. Where the winters are warm, such as near the Mediterranean, broad-leaved plants might retain their leaves in winter – i.e. evergreen oak and cypress trees – but in colder latitudes trees tend to be deciduous, shedding their leaves to reduce frost and wind damage. Conifers, with their thin needle-like leaves, are able to withstand very cold winters and remain evergreen, as can the tough and prickly holly.

Once our human ancestors roamed from the warm savannah of their origin, they would have needed clothes and shelter to keep them warm. The animals that they hunted as they moved north provided them with fur skins. Overhanging cliffs and caves were possibly the first shelters, probably followed by simple tents or tepees – which are not unlike an animal structure of bones covered with skin.[4] At first buildings were ephemeral and temporary because early humans were hunter-gatherers who followed the migrating herds. An early trace of human settlement is the Acheulian hut constructed some 350,000 years ago (Fig. 2.2). In geological terms this was the Pleistocene era, a period culminating in a series of ice ages in Europe. Retreating glaciers, swollen rivers and wind erosion had left hollows scoured into the softer bands of rocks on the sides of valleys. The combination of solar radiation and protection from cold north winds led to these sites, off the valley floor and south facing, being chosen wherever possible to take advantage of the micro-climate. The Acheulian hut was little more than a screen of branches laid against overhanging rocks: the site shaping the shelter.

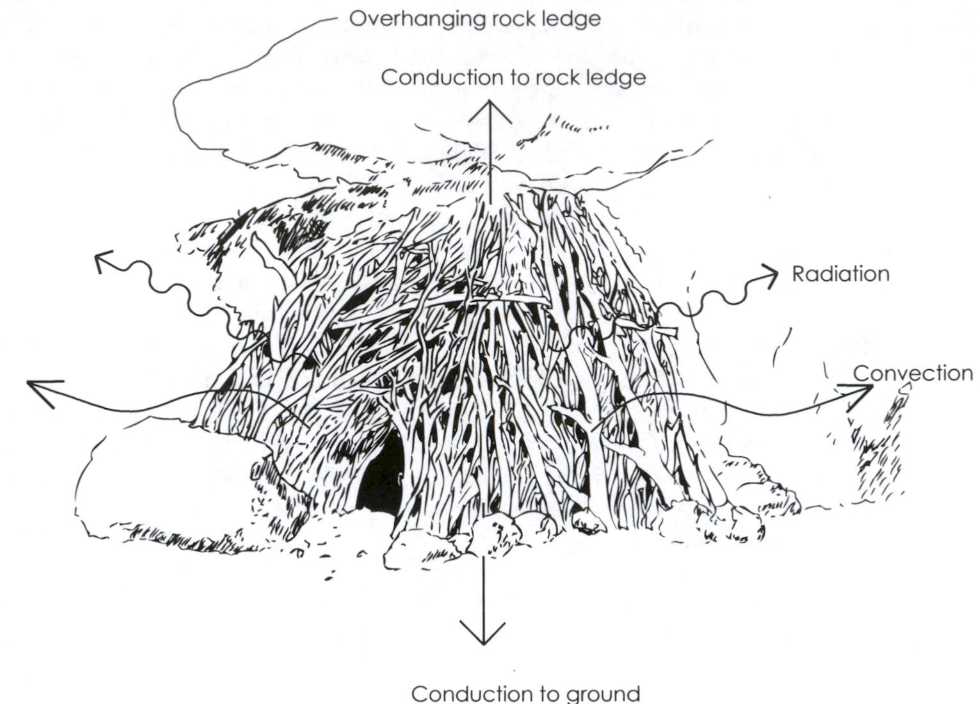

2.2 The Acheulian hut.

The occupants had fire, useful to ward off cold but also bears and wolves that found such sheltered places equally desirable. Fire came to be not only a practical matter of warmth and cooking but also a significant focus of what it might be to be human – adding to, or perhaps preceding, language, tool use and symbolic activity (such as rites of burial for the dead). It extended the hours of light in the long winter months, it gathered people in a circle, it would have prompted talk, and it would have formed a powerful image of 'home' in the memory of the wandering hunters.[5] Fire perhaps encouraged reflection upon the activities of hunting. Here, in the warmth and light at the centre of built shelter, possibly lies the origin of art and the qualitative aspects of human life. It is hard for us to imagine the significance of fire in our world of electric light at the flick of a switch, TV and electronic music. But as late as the mid-twentieth century, Edwin Muir could write of gathering around the fire and lamp in an Orkney croft listening in wonder to his father's stories.[6] Myths probably began in this way, stories weaving together explanation and imagination in a poetic form. And the famous Magdelanian cave paintings provide early evidence of art, extraordinarily detailed observations of animal life recalled and recorded in the light from fire.

Sites and climate

Woven fish traps, windbreaks and baskets provide some of the earliest evidence of man-made things.[7] The architect and theorist Gottfried Semper suggested that weaving was the first act of making and that the partition and the hearth were the primary archetypes of dwelling. People who adapted to hot, dry climates, such as native Australians, often did not build houses but only windbreaks. Sitting on the beach sheltered by our windbreaks today we feel not only warmer but also more secure having marked out a territory that temporarily belongs to us. Centre and boundary remained fundamental to the development of many different architectural theories up until the Modern Movement in the early twentieth century, which aimed to dissolve the distinction between inside and outside, amongst other things. In most climates buildings are necessary to make life comfortable, in extreme ones they are absolutely essential for survival. In this section we will look at some examples in extreme climates to set the scene for what might be achieved by building (almost) alone.

The Arctic has one of the most extreme climates, but low temperature is no great drawback to cold-blooded creatures such as fish. Oxygen dissolves more freely in cold water and if a cold-blooded creature is cold then it metabolises less rapidly than when warm and so needs less energy, which means less food.[8] Warm-blooded creatures of the Arctic regions, such as seals, penguins and polar bears evolved layers of fat or blubber to insulate themselves, and reduced the blood flow to the surface. Humans, however, had to adapt by their own handiwork.

Having lived in the Arctic amongst the Inuit, the anthropologist Hugh Brody has recently upheld their hunter-gatherer lifestyle as a paradigm of how we should deal with the Earth. In this region dominated by snow, ice and bitter winds, they exploit only locally available resources: fish and seals, and caribou in the summer. The Inuit occupy an extensive territory mainly north of the 12°C August isotherm (which marks the tree-line) from Labrador to Alaska. Brody's is one voice among many calling for a change in the way Western energy profligate society treats the environment.

Unlike many nomadic peoples – most famously the Bedouin with their tents and the Mongols with their yurts – the Inuit do not carry their dwellings with them, but utilise the one readily available material: snow. The Central Inuit who live in the vicinity of the Hudson Bay adjacent to the Arctic Ocean spend the winter in igloos. The igloo is a very appropriate adaptation, made from compacted snow as found, cut into blocks and built up in courses to make a hemispherical dome.[9] A 5 m circle of blocks is laid in an incline and leaning slightly inwards so that as construction spirals upwards the dome needs no support to prevent it collapsing.[10] This is a logical and economical form of construction and also a good shape to minimise resistance to the fierce winds.

Entry is through a curved tunnel cut into the snow, which helps keep out the wind, covered with a barrel vault of snow blocks. The entrance is often on the windward side, for snowdrifts can pile up in the lee of the wind. A low, snow-block wall helps to prevent wind entering (Figs 2.3 and 5.2). Whenever possible the Inuit choose sites for their igloos under the shelter of cliffs or on high ground at the edge of the sea to give protection from wind. Skins are pegged to the inside of the igloo forming an insulating layer of trapped air, not unlike in an animal's fur. Furs also provide material for the Inuits' clothing.

In winning the race to the South Pole in 1911–12, Roald Amundsen drew upon two

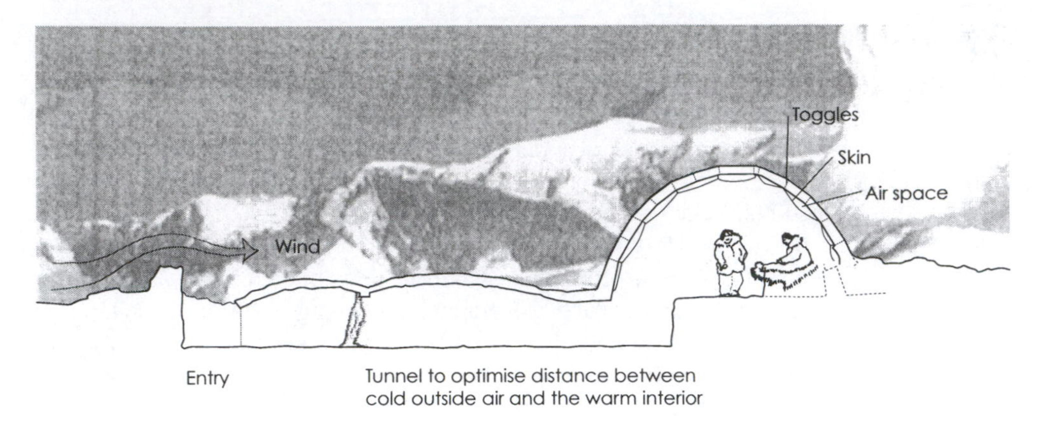

2.3 *Section of an igloo of the type used by the Inuit of northern Canada (after Rapoport).*

years he had spent studying the Inuit way of life and methods of surviving in the extreme cold. He adopted native dress and went with teams of dogs pulling sledges. In contrast, Robert Scott followed the tradition of the Royal Navy in asking his men to pull the sledges. Scott took three motor sledges and 17 ponies, which became encrusted in frozen sweat and had to be put down (dogs sweat through their tongue and hence are not affected in this way).[11] Adapting to the climatic conditions in traditional, native ways Amundsen survived, whereas Scott and his companions died in their attempt, relying on Western clothing and technical means. A recurring theme of what follows will be the valuable lessons to be gathered from the ways traditional societies and their material cultures responded to the environments they inhabited.

In the other extreme climatic region – the desert – plants and animals have evolved strategies for surviving the fierce heat. Cacti are the best known examples. They have evolved to exist without leaves, thus reducing water loss by transpiration, and their stems are swollen with stored water. To reduce moisture loss during the heat of the day they use a special process of photosynthesis by which the stomata are opened at night to take in carbon dioxide, which is stored within the tissues until it can be photosynthesised during sunlight hours. Mammals have adapted to the desert in a related way by staying in burrows during the day to avoid the sun, emerging only in the cooler night air, unlike cold-blooded reptiles that get some of their energy from sitting in the sun.

Like the burrows of desert animals, the Matmata dwellings of southern Tunisia at the edge of the Sahara are dug deep into the earth to make use of its thermal capacity (i.e. its ability to store heat and coolth) and help provide a stable temperature, in this case for cooling.[12] A major factor here is the extreme diurnal temperature range as well as the sun's altitude and intensity. A sunken courtyard forms the centre of the Matmata dwelling around which rooms are carved out of the surrounding soft limestone (Fig. 2.4). Recent environmental research has shown how effective they are in modifying the climate.[13] The graph in Figure 2.5 demonstrates how a stable and comfortable temperature is maintained in the rooms throughout the day and night despite the extreme diurnal range. Similar sunken courtyard dwellings exist in Rijban, southern Libya. On the North China Plain dwellings have been dug into the soft loessial soil stabilising room temperature at 11–20°C when ambient winter temperature falls to 0°C and in summer reaches 30°C.[14] Like many vernacular forms of building, the Matmata

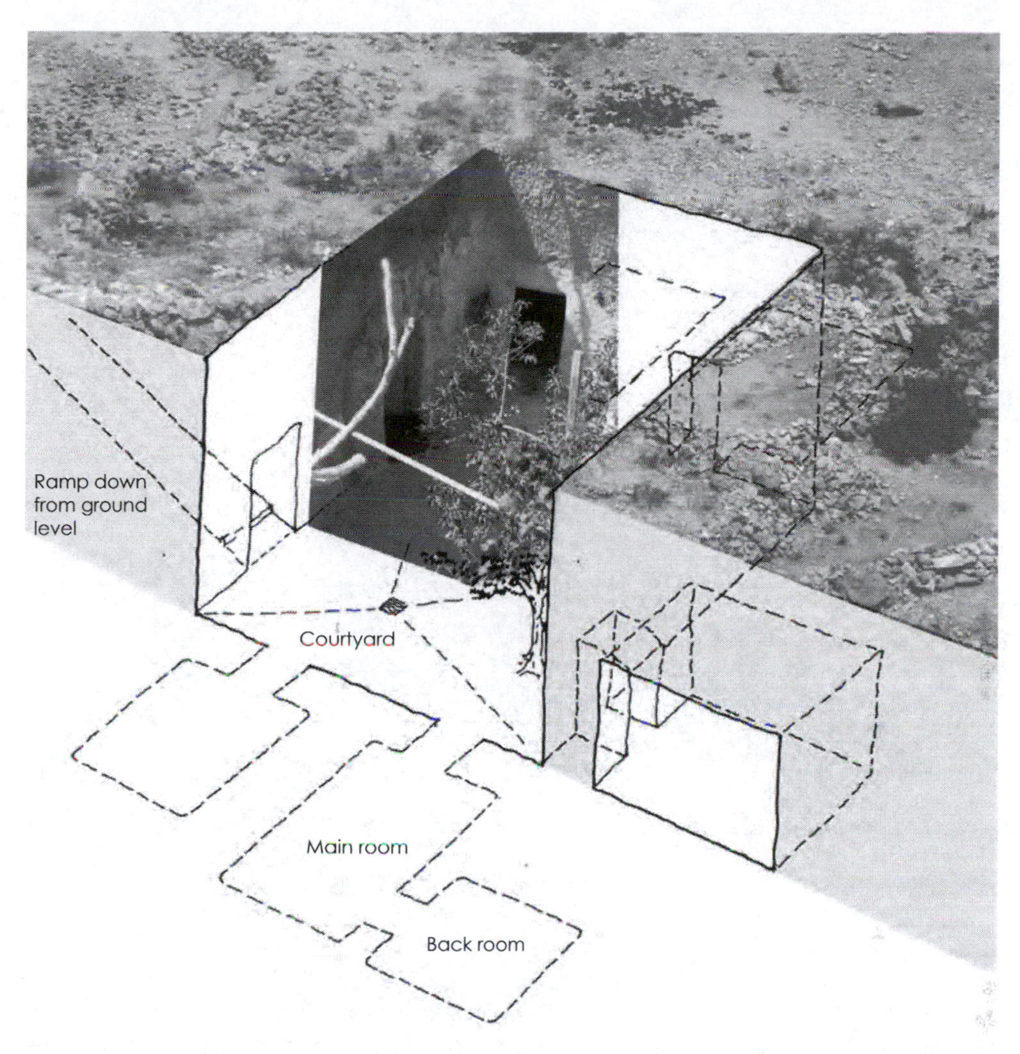

2.4 Axonometric of Matmata dwelling. The photograph in montage is of a similar house in Rijban, Libya.

has fallen out of use under the attractions of a modern lifestyle in modern surroundings made possible by air-conditioning. (However, there is a growing interest in using the Tufa caves in some places such as the Loire Valley in France.)

These examples are hardly buildings at all, being little more than hollows in the surface of the Earth. But the ingenuity of vernacular buildings in hot climates always astonishes. In the hot desert climate of Iran, for example, there are not only the evocative 'wind-catchers' for cooling houses (discussed in Chapter 6), but also ice-houses. A thick wall built on the east side of a diverted stream helps still the air, thus allowing the water to freeze on cold nights and casts a shadow over the thin ice at first light. The ice was collected and stored in thick, circular buildings attached to the wall. In one example building design exploits air movement, in the other it prevents it – different strategies for different purposes.

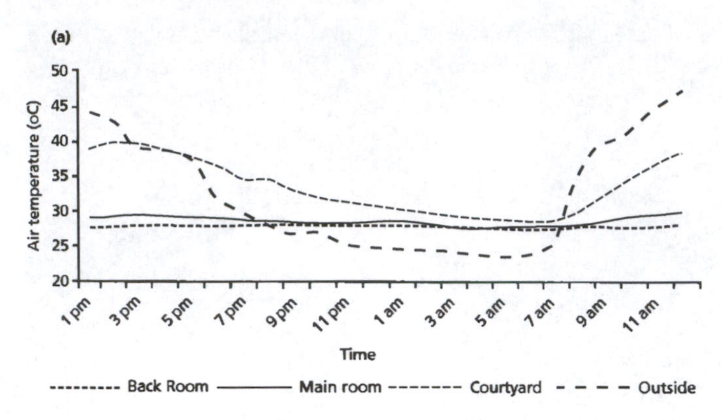

2.5 Graph showing temperature in the Matmata dwelling. Temperature recorded at Chenini, Tunisia on 11–12 August 2000.

2.6 Marsh Arab dwelling in southern Iraq (drawn by Bill Ungless).

In the extremely hot flood plains at the mouth of the Tigris and Euphrates, the Marsh Arabs of Iraq used the one readily available material for their dwellings: reeds. Bundled together and bent over in graceful hoops, they formed great ribs, like a barrel vault, which gave structural strength and also some thermal mass to absorb some of the sun's heat. Between the ribs were open-weave reed mats that allowed air to pass freely through (Fig. 2.6). Tragically, Saddam Hussein drained the marshes after the Iran–Iraq War. However, efforts are now being made to restore the original watery landscape.

These dwellings are types often described by anthropologists as central to particular cultures because they continued in use for very long periods of time due to their successful adaptation to the climatic conditions. They made use of the limited range of material available for building in their region and they developed strategies to make comfortable living conditions in temperatures otherwise intolerable for humans.

Farming and the landscape

As hunter-gatherers, earlier mankind had little impact upon the broader environment. But with the 'invention' of farming this changed. Farming began sometime between 10,000 and 6,500 BC in the Middle East and a new world-view prevailed. Humankind entered into what has been called a 'social relationship' with plants and animals in the domestication of wild species.[15] The title of Brody's book, *The Other Side of Eden*, indicates that he sees farming as a 'Fall' from grace. Farming allowed settlement and the accumulation of material artifacts. Over many generations settled people developed a way of life and material culture that of necessity drew upon the resources available in their immediate environment. The first farmers had to rely on natural sources of energy – the sun's warmth, trees for firewood or for shade – and locally available materials for building. The limited evidence from some of the first Neolithic farmhouses shows they responded to climatic conditions. Linearbandkeramic (LBK) 'longhouses' along the Danube, for example, had a narrow closed rear end facing the prevailing wind and possibly an open porch on the leeward end[16] (Fig. 2.7). The frame would have been filled with woven panels of wattle and covered with a daub of clay, similar to much later medieval half-timbered houses.

Humankind developed a new relation to the land and the processes of nature over which they gained increasing control. This is sometimes called 'the Neolithic Revolution' and the patterns farming made upon the land and human culture continue to greatly affect us today. Farming needs enclosures for animals and fields for crops. A widespread desire of many people in the West remains to own a house on their own

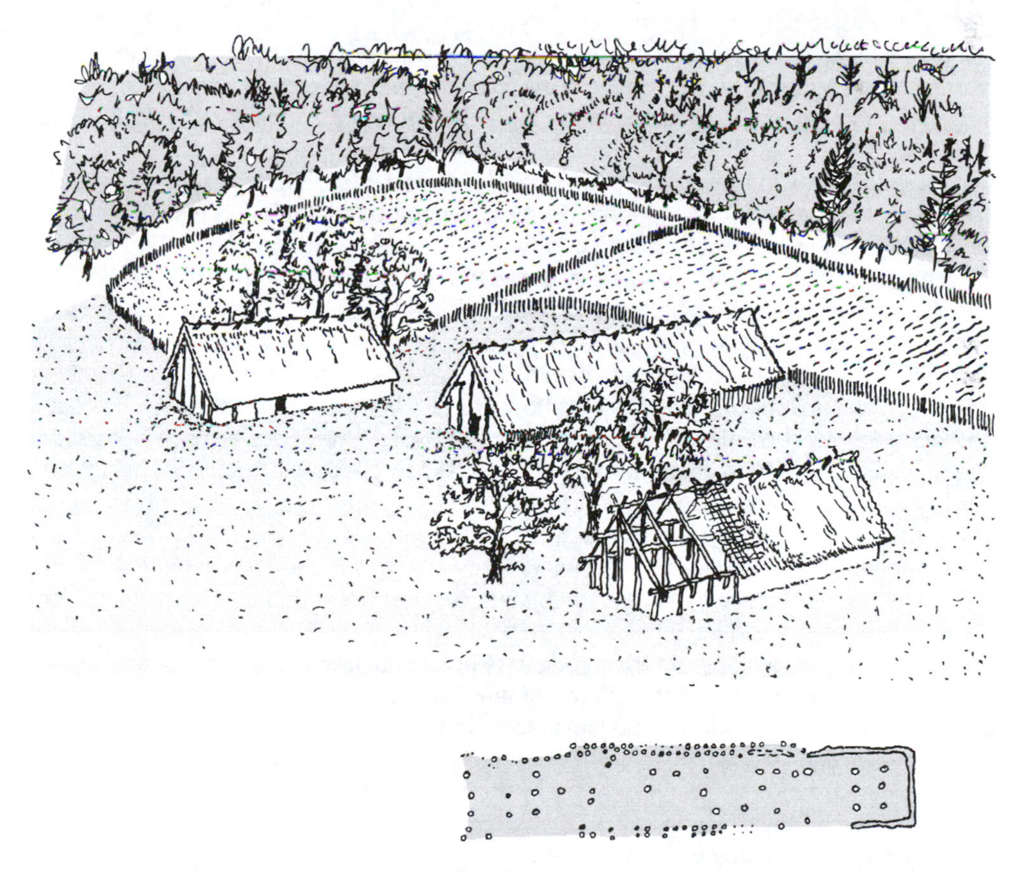

2.7 Reconstruction of Linearbandkeramic (LBK) longhouse of the kind built by Neolithic farmers as they settled beside the Danube. Based on plan below as excavated by archaeologists.

patch of land, an echo perhaps of our original habitat. The lingering memory or myth of the independent yeoman's farmhouse in particular played a part in promoting streets of cottage-like suburban houses when the retreat from the city began during the Industrial Revolution.

Communities of living things change as time passes. There is a pattern in all kinds of environments where new species enter the community and displace others. This parade of species is called succession and eventually comes to a halt or 'climax' when vegetation and animals reach equilibrium, replacing and replenishing themselves. In temperate lands such as Britain, the natural climax is oak forest. The first farmers to arrive in Britain would have found it covered by an almost impenetrable forest dominated by oak in which they made clearings for fields.[17]

In the Orkney Isles, off the north of Scotland, the extraordinary Neolithic settlement of Skara Brae remains nearly as complete as when built 5,000 years ago. Buried in sand for 4,500 years, it emerged to illustrate how early humans adapted shelter to particular environments and climate, building working together with site to make human existence possible even in the most inauspicious conditions. Orkney's indigenous scrub of hazel, willow and birch was quickly cleared by the first settlers and never regenerated, partly because of the frequent gale force winds and salt-laden air.[18] Because the islands are virtually treeless, the houses are built of stone – hence their remarkable state of preservation.

Like all of northwest Britain, Orkney has its climate modified by the Gulf Stream – a warm current transporting water from the Equator northwards. Skara Brae is further north than the Hudson Bay or Moscow, and environmental conditions would be much colder were it not for the moderating influence of the sea. Long cold winters and icebergs, like off Newfoundland, are predicted for Britain if the Gulf Stream were to disappear, or be diverted, due to global warming, as some research suggests it might.

A combination of often low temperatures and high winds (characterised by the wind-chill factor) and rain driven by frequent Atlantic gales were the principal climatic conditions that Skara Brae had to deal with. The dwellings were sunk partially into the earth, minimising contact with air and wind as well as making use of the thermal capacity of the ground, in this case obviously for warmth rather than coolth. (Evidence from the Czech Republic and Russia show Upper Paleolithic houses using the same strategy of sinking dwellings into the ground.[19]) Skara Brae houses have thick stone cavity walls with a high thermal mass – the cavities filled with 'midden' (weathered domestic waste, like compost) – and the eight or ten dwellings share an underground entrance passage that would have acted as a draught lobby[20] (Fig. 2.8). The Koryaks of Northeast Arctic Asia lived in semi-dugout dwellings into the twentieth century, and the Ainu (early hunter-gatherer inhabitants of northern Japan) built winter 'pit dwellings' covered with earth.[21]

The underlying geology of Orkney, distinct horizontal layers of flagstones and sandstone, is revealed all around the shores of the many islands that make up the archipelago. Exposed in low cliffs, the laminated beds suggest a natural model for the dry-stone walling of the houses. Semi-submerged in the ground, these early stone houses respond directly to the geology of the place. Like the Acheulian hut, the igloo and the Matmata dwelling, Skara Brae appears little more than a slight modification of the landscape and hardly distinct from it.

Furniture at Skara Brae was made from the easily split local sandstone allowing us to see in great detail how comfortable these earliest houses were. Stone-walled beds

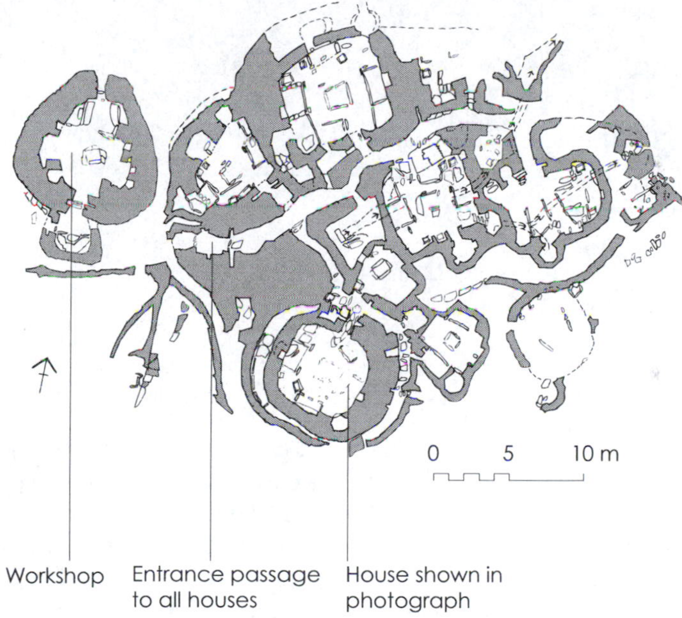

2.8 Plan of Neolithic settlement at Skara Brae, Orkney with house no. 7 shown above.

were probably filled with straw, chaff or heather for comfort and insulation and incorporate vertical stone pillars at the corners on which may have hung skins (giving the appearance of a four-poster bed, see Shakespeare's house in Figure 5.4). This arrangement would keep the bedding dry if the roof leaked.[22]

Looking down into these ancient houses evokes the snug and practical comfort provided by the building strategies. But there may have been a symbolic counterpart to the practical advantage of sinking the houses into the earth. A similar settlement at Rinyo on the Orkney island of Rousay, built hard up against a sandstone cliff-face to protect it from the north winds, has a complex system of drains to carry off rainwater

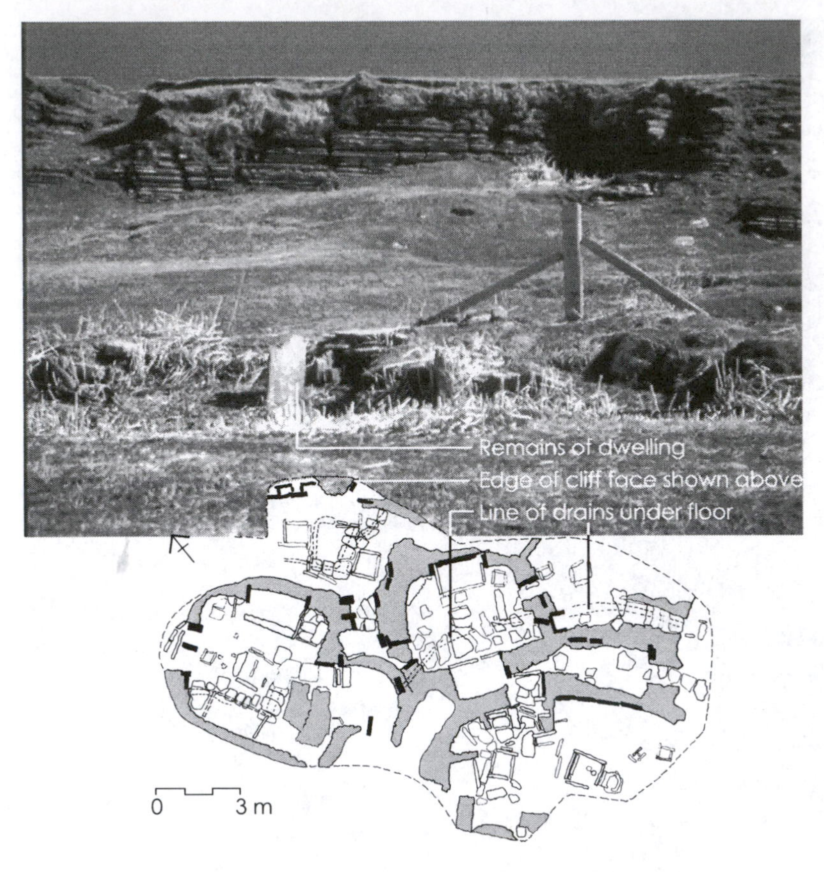

2.9 Plan of Neolithic settlement at Rinyo, Orkney, with fragmentary remains shown above.

that pours from the cliff (Fig. 2.9). Moving the house walls a metre away would have avoided this while preserving the microclimatic advantages provided by the cliff. It seems likely that the symbolism of building into the earth was paramount, tapping into the life-force that produced plants from seeds as they lie buried in its darkness. This might explain why midden or compost was piled around the stone walls and in the cavities, as at Skara Brae. Perhaps encasing the walls with earth was considered an essential symbolic practice because at this time building with stone was only for burial tombs.

In his seminal work *House Form and Culture*, the anthropologist Amos Rapoport reminds us that vernacular dwellings were not merely a response to purely physical needs. 'The great variety of forms, strongly suggest that it is not site, climate, or materials that determine either the way of life or (human) habitat.'[23] He illustrates the range of buildings that can exist in any one climatic region to challenge what he calls 'the physical determinist view'. He refers to Lewis Mumford's argument in his book *Art and Technics* 'that man was a symbol-making animal before he was a tool-making animal'.[24] The evidence for whether mankind was first of all a tool user or symbol maker is problematic, but nevertheless anthropologists and archaeologists generally agree that symbolic thought and behaviour are as important ingredients as responding to physical needs in the formation of distinctive cultures. It is vital that this be kept in mind when developing a sustainable architecture; climate, culture and place go together. These are themes we will return to in the following chapters.

Most landscapes have been shaped by thousands of years of farming and are characterised by a combination of topography, vegetation and buildings. Although cultural and symbolic factors played an important part, nevertheless traditional or vernacular buildings responded to the environmental conditions where they were situated, as do plants. Often the buildings are so well adapted that they seem to have grown from the earth. These vernacular buildings record generations of careful adjustment to site and climate, to the materials available and particular weather conditions. In northern temperate, rainy climates, roofs tend to be steep, for example, whereas in mountainous areas with predictable and heavy winter snow, roofs are less steeply pitched so that snow can settle and provide insulation. In both cases south-facing slopes are chosen to maximise solar gain, as did the Acheulian hut. In the hot south, roofs are usually flat, as there is little rain. Both the roof and walls are built solid and thick to reduce heat penetration. In hot but humid tropical regions, roofs are very steep to throw off the frequent heavy rain and walls are a flimsy screen to allow maximum air movement through the building. The size and number of windows is reduced in buildings nearer the Equator: the sun excluded in the hot south but invited in further north.

The Earth's underlying geology not only shapes the pattern of land-use and built form, but also provides material for building. Alec Clifton-Taylor's book *The Pattern of English Building* is a beautiful portrayal of how the vernacular buildings of even slightly different regions have found their evocative form in response to local material and climate. If the proverbial man from Mars had studied Earth's geology, for example, he would be able to locate where he had landed in England if armed with this book. The vernacular buildings of a region in which a new design is to be situated are always worth studying as they provide a body of tried and tested knowledge about appropriate materials and forms. The habitual practice of craftsmen made the places of human inhabitation. Both words – habitual and inhabitation – are derived from the Latin *habitare*, to dwell, as does habitat and habit – an old English word for clothing (and one still used currently in French). The vernacular grew from small adjustments to strictly observed craft traditions that produced built forms passed on from generation to generation. They demonstrate the traditional role of buildings as passive modifiers of the environment, or 'as thermal control devices' making comfortable living conditions in all kinds of climates.[25]

The Industrial Revolution did much to destroy the integral relationship that had existed between vernacular buildings and their immediate environment, with building materials being drawn from the ground on which they sat or from the surrounding fields and forest. New means of transport meant that materials, for example, could be imported from distant sources and usually were, if cheaper, as was the case in nineteenth-century Britain with ubiquitous Welsh roofing slates. The enormous expansion of cities and the new inventions of gas lights and electricity followed by mechanical ventilation and air conditioning, saw buildings and their inhabitants become increasingly independent of their 'natural' environment and climate.

The machine in the landscape

The systematic organisation of science in the Enlightenment subjected traditional forms of knowledge to the scrutiny of reason and eventually overthrew traditional ways of doing things. The precepts of scientific enquiry and a parallel belief in progress eventually encouraged the primacy of new invention over time-honoured tradition. With the steam engine, the Industrial Revolution discovered an immense new source of power, 'liberating' humans from the bonds of nature. Hitherto mankind's own muscular power was only extended by the bend of a bow, the strength of domesticated animals, the wind to drive boats, water and wind to turn mills. The rise of the Romantic Movement at the end of the eighteenth century was in part a critique of burgeoning industrialisation. Poets such as Wordsworth and Coleridge extolled the virtues of all kinds of traditional labour, seeing in these more appropriate ways humans should deal with nature. Romanticism stressed involvement in the natural world rather than the objective detachment of science.

We find ourselves using the words poetry and poetic all too frequently for the elusive qualities of space and form that elevates prosaic building to architecture. Like the old adage 'beauty is in the eye of the beholder', this can be vague, purely subjective and meaningless. In general terms, poetry might be considered as the selection and arrangement of words that offer a fresh view on familiar things, phenomena or emotions. The work of Elizabeth Bishop springs to mind here, continuing a tradition that perhaps began with the Romantic poets.[26] The way poetic is used here alludes to this sense of the word but also echoes how the philosophers Martin Heidegger and Gaston Bachelard have discussed it. Heidegger writes of our poetic relationship to the world through making, and in his book *The Poetics of Space* Bachelard discusses how spatial characteristics inform our relationship to our inner selves. Bachelard says that the house is 'the human being's first world' and that 'man is laid in the cradle of the house'.[27] He discusses how certain kinds of spaces – their shape, size, arrangement and structure – resonate with the full depth of the developed psyche and allows for 'well-being'. We will return to this at the beginning of the next chapter, which deals with houses.

In *The Question Concerning Technology*, Heidegger shows how the original meaning of the Greek word *poesis* was linked to the word *phusis* – from which our word physics is derived.[28] *Phusis* referred to things coming out of themselves, as plants grow from their own seed. *Poesis* resonated with this in that it meant a form of making material things that fitted the harmonious order of the given world or cosmos. Heidegger demonstrates a close etymological relationship between *poesis* and *techne* – from where our words technical and technology are derived. His overall aim was to show that, in its origin, technology was a form of carefully fitting into the world rather than aggressively taking from it. At their best, vernacular buildings confirm Heidegger's view.

The Arts and Crafts – which can be considered an offshoot of Romanticism – was largely concerned with resisting the impact of industrialisation at the end of the nineteenth century. An extreme example is Edward Prior's 'Home Place' (1903–05) in Norfolk. The house is rooted to its site, quite literally, for the sand and gravel aggregate for the mass concrete walls and floors (reinforced with chains) came from a sunken garden dug out of the ground as part of the landscape design (Fig. 2.10). Larger stones dug up were used for facing the building and clay tiles were locally produced. Unlike many of his more whimsical peers, Prior was a rational Arts and Crafts architect. But although extremely modern in its use of concrete, the expression of the building is traditional and its response to place and site was a legacy of the Romantic Movement's desire to locate man closer to nature. Prior designed the 'butterfly' plan

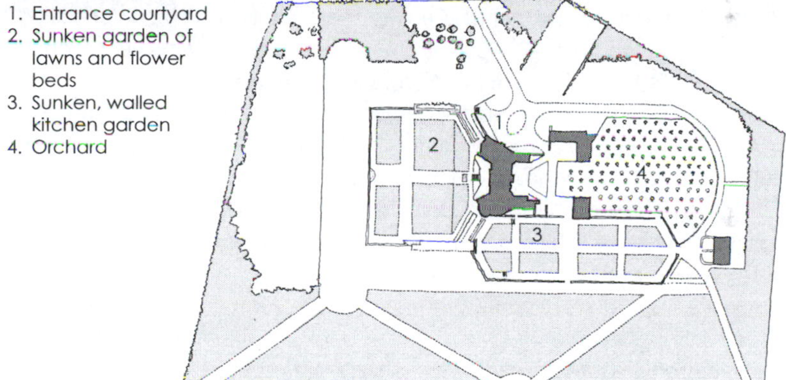

2.10 Home Place (formerly Kelling Place), Holt, Norfolk designed by Edward Prior.

to open more rooms to the sun. This drew parts of the site into the 'arms' of the building and also had the effect of breaking up its form. A general lesson for a sustainable (or any) architecture is that a building can play a part in shaping the site, not simply be a shape on the site. For many designers today a site's geology, topography, orientation and vegetation will suggest ideas of architectural form and appearance, in the largest sense.

It became clear to the pioneers of modernism, however, that the twentieth century was to be industrial and they believed that architecture should reflect this and become machine-like. Le Corbusier's Villa Savoye epitomises this, its pure white forms floating like some unearthly machine poised for lift-off. Cars drive through an Arcadian landscape, park under the house, and visitors walk up a ramp to the living area where they

2.11 Perspective sketch of the Villa Savoye (after Le Corbusier).

can look out over the landscape through slit windows (Fig. 2.11; see also Figures 3.16 for photograph of the exterior and 7.15 for an interior view). Engineered forms, such as cars and airplanes, he believed to be evolving as efficiently and beautifully as natural selection produces plant and animal forms. The title of Sigfried Giedion's book, *Mechanisation takes Command*, concisely summed up one aim of this phase of architecture in the innocent age of the machine, when energy was cheap and before the impact of industrialisation upon the environment was detected. But, as Hassan Fathy has eloquently said, 'A machine is independent of its environment. It is little affected by climate and not at all by society.'[29] One aim here is to consider the design of buildings, not as machine-like and independent, but in more biological terms, as open to their environment, site and climate, as plants and animals are to their habitats. In polluted cities this may be less possible, and some buildings may need to be as defensive as cacti, for example, in developing strategies for an inhospitable environment.

Shortly after the Villa Savoye, Le Corbusier designed another house with an approach to site planning offering more fruitful lessons for us today. His drawing for the Petite Maison de Weekend (1934–35) shows a line across the site beyond which the car cannot pass (Fig. 2.12). The parked car is contained by an earth-bank that rises up to become a turf roof on the house. The house is thus rooted to the ground and opens up to a site structured to provide graded relationships between interior and exterior, house and garden. Although his earlier work advocated the idea of buildings as machine-like vessels – albeit often incorporating the pleasures of ship-like decks – his later buildings became structured by the reciprocity between light and shade and open themselves up to the environment. Le Corbusier was a complex figure to whom we will return.

In a small and modest way, needless to say, the Petite Maison de Weekend recapitulates those grand, late Renaissance villas, such as the Villa Lante or Villa Aldobrandini, that laid out the site in a graded and ordered sequence from 'wild' nature beyond the bounds to civilised order at the centre of the house. The eighteenth-century gardens at Chiswick House (designed by Lord Burlington and William Kent) were structured by this idea and incorporated orange trees in theatrical settings to promote the idea of mankind's role as bringing nature's latent perfection to fulfillment. Oranges – looking like balls of gold, and the tree's blossom smelling like paradise – could only survive in these northern climes thanks to the tending of gardeners in conjunction with the technical development of greenhouses. The idea of the man of virtue, or 'virtuoso', was promoted in opposition to scientists – newly emerging at this time – who were bidden to put 'nature to the rack to reveal her truths' by Francis Bacon (1561–1626), one of the founding fathers of science.[30]

The ideas about mankind's relation to nature expressed in these gardens contrast

2.12 Axonometric of the Petite Maison de Weekend (after Le Corbusier).

greatly with the prevailing economic model of our own consumer society that sees nature as a stockpile of resources. Sites too tend to be seen as economic resources to be maximised rather than primarily in ecological terms as part of an ecosystem into which they need to adapt. In touching upon the early critique of science at Chiswick, let us reiterate that science should not be confused with technology. Science deepens our understanding whereas its application through technology is shaped by the prevailing world-view – in our time dominated by the economics of consumerism. The examples we have seen from the passive past provide us with lessons for how a site might be treated, how buildings can best fit a site in any particular area, what material might be suitable, an appropriate scale, etc.

Site strategies

Many plants evolved to disperse their seeds in a variety of ways and take root and grow where the conditions are propitious. In the early days of mankind's occupation of the Earth, buildings could find an optimum place and adapt to it like an animal or take root like a plant. Populations were much smaller, there was more open land, and people could choose the best place to site a building. In contrast, today sites are largely determined by other factors including economic, the density of population, shortage of land, etc. Any given site will be in a particular climate zone, which will be a primary, if general, consideration. Most climate zones will have different conditions between summer and winter, which will entail a building requiring some flexibility of response.

A good starting point is to be clear about the fundamental characteristics of the climate. The UK, for example, is divided into two climate zones, as we have seen – 'Mid European coastal' in the south and 'North European coastal' further north. Low solar radiation and mild summers characterise both zones with the south having cool winters and the north cold winters. Traditional buildings in most of the UK have therefore been principally concerned with protecting occupants from the 'cold, wind and wet of the relatively long cool season'.[31] (Skara Brae was only an extreme and early typical response.) The combination of global warming and thin-skinned, highly glazed modern building has introduced summer cooling as a major factor in temperate climates such as the UK. Consequently each of the environmental factors associated with any site in any region – sun, wind, rain and the ground itself – needs to be considered in a rounded way, taking into consideration the two extreme seasons of winter and summer. This could be considered as the 'site from the environmental point of view'. Figure 2.13 shows some general characteristics of a site and some of the environmental considerations that will be outlined below.

Solar considerations

Orientating the building to maximise solar gain can reduce the amount of energy required for heating. Similarly the position and size of windows can be designed in response to this. The sun appears to move around the site from east to west during each day. But the sun's trajectory over the Earth changes from winter to summer. In

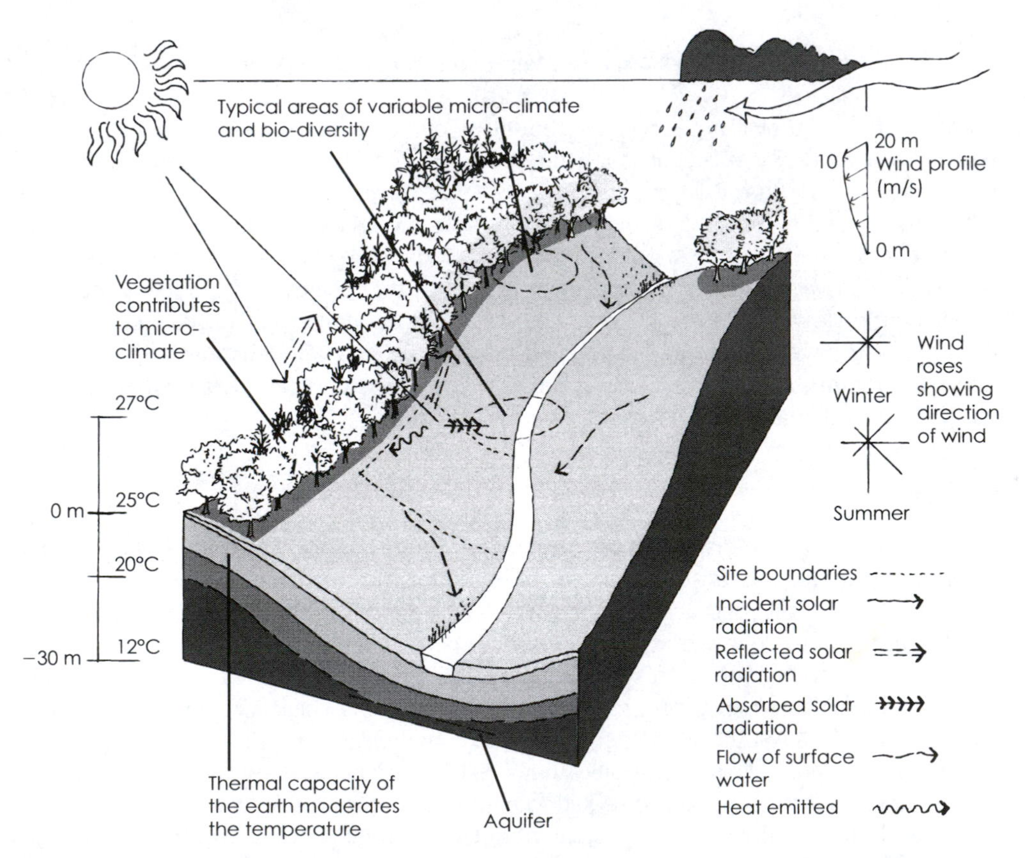

2.13 Diagram showing some general characteristics of a site and environmental considerations.

winter in the Northern Hemisphere it will rise south of east and set south of west. In summer it rises north of east and sets north of west. Although the sun is always at its highest at noon, in mid-summer its altitude at the Greenwich meridian is about 62° and in winter it rises only to 14°. The solar path diagram shown earlier in Figure 2.1 is a very useful tool to understand how the sun moves, rises and falls, and helps locate a building on a site sensitively in relation to potential solar gain.

To maximise solar gain, south-facing windows need to have nothing on the site that casts shadows upon them. The most effective distance to avoid shadows between the building and any neighbouring one can also be calculated. However, solar gain can also be excessive in summer, so some shading may be beneficial. In this case locating the building near any trees that exist on the site, or planting new ones could be considered. Deciduous trees at the right distance from the building could give shade in summer yet allow some solar gain in winter. Design can involve difficult choices in relation to competing demands and, as is often the case, the successful solution is the one that achieves the right balance.

These factors, of course, relate to passive solar gain in reducing the heating load on a building and controlling the cooling load in summer. But with solar panels and photovoltaic cells making it possible for a building to actively generate energy, it is important to consider how these elements are integrated with the building's design and siting to maximise their potential. South-facing, unshadowed roofs are the best locations but south-facing walls have some potential.

Window openings not only provide solar gain (and potential heat loss) but also views and natural light. Good design simultaneously considers the effects of the outside on the inside and, similarly, how the internal environmental considerations will affect the external appearance. Light comes from the sky, so overshadowing should be avoided. Views are for the eye, so if there are good views within or beyond the site, the planning of the building and its openings should be carefully considered in relation to these, and the site laid out accordingly.

When a building is in place it will affect the microclimate of the site around it. For example, the environment adjacent to the south wall will be markedly different from the north. Specific varieties of plants will wither, survive or thrive in these different conditions of warm sun or constant shade respectively. Plants are not only adapted to a particular climate but also modify the microclimate itself. Vegetation takes carbon dioxide from the air and produces oxygen; it can absorb some noise and raise local humidity. Perhaps most importantly, plants add to the general quality of life, not only making a site more pleasant for humans but also habitable for animals, birds and insects. A site's contribution to the biodiversity of an area will be assisted if the plants chosen are diverse and some are native to the region.

Many modern building types – supermarkets, factories, air and rail terminals, etc. – have a very big physical 'footprint' (the concept of 'ecological footprint' will be discussed in Chapter 8). Consequently these take up large areas of what are existing or potential habitat of other creatures. Green roofs on these and other buildings can be designed to replace such lost habitats. Buildings can provide huge opportunities for biodiversity, states a report produced by the Construction Industry Research Information Association: 'Why can't we create limestone grassland on roofs, or any other habitat?'[32] Inspired by a meadow that grew on the roof of a water filtration plant in Zurich, one author of the report has built a green roof on a Waitrose store in London's Docklands. This has become the habitat of a black redstart, a bird that once thrived on rubble-covered demolished building sites. In this way a building itself can begin to replicate the ecosystem its 'footprint' has taken, or even make a new kind of habitat.

Walls can also be the 'site' for plants. Climbers have traditionally been used for centuries, but modern methods using steel straining wires have expanded the possibilities. The foliage of creepers provides effective habitats for birds to nest, displays seasonal changes of colour, as well as offering some protection from sun and wind.

Wind

Prevailing wind directions for a particular region can be found from wind rose diagrams that are available from the Meteorological Office (Figure 2.13 shows an example). These can be used to find the dominant direction of the wind in winter and summer. A building's energy consumption can be reduced by restricting wind in winter, but a breeze assists cooling in summer. Particular sites may have local factors that affect this and these need to be ascertained and inform the design of a building and its siting. Making use of existing vegetation or planting trees and bushes can form a windbreak to reduce wind speed and the resultant heat loss. The sound of wind in the trees can add a sense of animation to a place in addition to creating zones of privacy. Planting trees often has the advantages of adding beauty and encouraging biodiversity. Over the life of trees the process tends to be carbon neutral, as the carbon taken up is eventually released when the tree dies and decays. The long life of trees, however, endows them with a durability and persistence that is to be valued in a throw-away, consumer culture. Timber can also, of course, be used as building material and re-cycled. Fast-growing trees, such as willow, can be incorporated in local biodiversity projects.

The Australian architect Glenn Murcutt has designed the Henric Nicholas Farmhouse (1977–80) to combat cold local winds in a related if more traditional way. He placed a pair of large water butts against the side of a very modern steel and glass farmhouse to act as a buffer or windbreak. A curving roof (perhaps echoing the crown of a tree) helps to deflect the wind over the dwelling at the same time as it invites a little sunlight into the kitchen. Murcutt describes this strategy as 'turning up a collar to the wind' (Fig. 2.14). His architecture can be seen as an attempt to forge a new vernacular out of the conditions and materials of modern life yet one responsive to place, site and climate. We will return to this theme.

Using the slope of the land to give protection from the wind is an established practice of vernacular dwellings all over the world from the Alps to the Arctic Circle. Open aspects of farmhouses often face down the valley and the eaves at the rear are low to the ground, as buildings have been partially dug into the hillside. Another contemporary architect who combines vernacular traditions developed in a mountainous region with modern ideas on form and making is Peter Zumthor. His Church of St Benedict (1987–89) is sited on the slopes of the Swiss Alps at Sumvigt. Its structure is made of softwood from the indigenous local pine and clad with timber shingles; it is designed in plan, almost like an airfoil, to offer least resistance to the occasional violent winds that sweep down the valley (Fig. 2.15).

The wind on a site can also be used actively to power wind turbines, which can be free-standing or attached to a building, large or small. These turbines can reduce the building's own energy demands on the grid as well as potentially providing electricity to it. Extensively used, micro-generators could help reduce the need for large-scale conventional generating facilities or wind farms in the country. If imaginatively incorporated

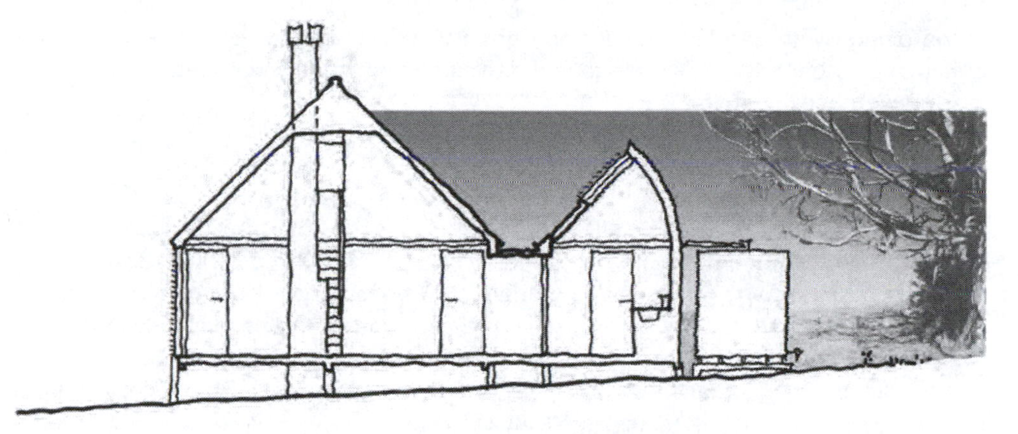

2.14 Section through the Henric Nicholas Farmhouse, Mt Irvine, Australia, by Glenn Murcutt.

2.15 St Benedict church, Sumvigt, Switzerland, by Peter Zumthor.

into a design with careful consideration given to scale and noise, perhaps the wind generators of the future will draw to themselves something of the romance we associate with vernacular windmills.

Rainwater

In northern climates a major functional role of buildings – both traditional and modern – is to protect the interior from rain penetration and thus the discomfort that would result and the damage to the building's fabric. A building prevents rain entering the ground directly and usually diverts water away through drains and sewers. Land drains (originally used on agricultural land) take water directly to ditches or rivers rather than allowing water to be absorbed slowly into the ground. Traditionally rain, a scarce resource, was often collected in water butts – as it is in Glenn Murcutt's design described above – and used for watering the garden or various domestic tasks. Rainwater can be collected and used as 'grey-water' for flushing toilets in the building. 'Grey-water' from washbasins and sinks can also be collected in ponds on site designed as reed-beds that act as a waste treatment system. Reed-bed filtration can reduce the impact on the local and regional drainage network.

Increased demand for water and its relative scarcity is a major concern. (The south-east of England has suffered its worst drought for 100 years at the time of writing.) We are drawing what has been described as 'fossil water' from aquifers at a much greater rate than it is being replenished, and imaginative ways of collecting, storing and reusing rainwater should form a key part of sustainable design.

Fast run-off of rainwater from buildings and land drains has been a significant contribution to widespread recent flooding in the UK. Hard, impervious surfaces used in landscaping the site add to this and site strategies that make use of the rainwater can help alleviate it. In addition to the possibility of creating habitats described earlier, a turf roof will absorb water, allowing some to evaporate, both reducing and slowing the rate of water going into the drainage system.

Site planning should aim to make use of rainwater, which could be collected in pools as part of the landscaping of the site. Evaporation from pools or fountains and the use of plants to moisten the air form an integral part of traditional cooling strategies in hot countries such as Spain and are often associated with courtyards. The Alhambra is a magnificent example of the use of water – although it receives this via channels from nearby mountains – with paths lined by channels of flowing water and courtyards with tranquil pools, fountains and plants (Fig. 2.16).

There is also a possible active use of water because hydropower can be generated if there is running water on the site. It is estimated that there are 20,000 disused watermills in the UK, some of which are now being converted for this new purpose. The Queen's residence at Windsor Castle is to be provided with electricity from four turbines built into Romney Wier on the Thames about 800 m away.[33]

The ground

As we note at the beginning of this chapter, a building becomes located in a particular place. It is rooted in a site through the practical need for foundations that allow the weight of the building to be carried by the ground. Often this entails a great mass of concrete and the removal of much earth from the site. Careful consideration should be

2.16 The Court of Myrtle in the Alhambra, Granada, Spain.

given to the foundation design to minimise its impact and a landscape strategy developed that makes use of the earth taken up on the site if possible.

A radical approach to foundation design is made in Penoyre and Prasad's Millennium Centre at Dagenham, a flagship ecological building that makes imaginative use of recycled materials (Fig. 2.17). The building is supported on stainless steel helical (corkscrew) foundations hence reducing the impact on the site itself and in addition they can be removed when the building's useful life is finished. 'Touch the Earth lightly' is an Australian Aboriginal saying that has been applied to the buildings of Glenn Murcutt, a phrase that is apt here and could be kept in mind for one approach to designing sustainable architecture. Pipes cast into pile foundations can draw upon the stable temperature of the ground for heating or cooling, as described below.

Some recent projects make use of the earth's thermal stability, as did the Matmata dwellings, but in a new, active sense. In Britain, for example, the ground temperature is higher than that of the air in winter and lower than it is in the summer, making it a potential source for both heating and cooling. Penoyre and Prasad's recent reworking of a building to provide a restaurant for the concert hall at Snape Maltings makes use of this principle in extracting ground water at a fairly constant 11°C to cool the internal environment, the system designed by the environmental engineers Max Fordham (Fig. 2.18). The addition combines traditional forms and innovative environmental engineering, continuing a long process at this site whereby vernacular traditions have been modified in response to new demands and new technologies. The original nineteenth-century buildings were adjustments of vernacular barns in response to the industrialisation of the malting process. When the Maltings fell into disuse in the mid-twentieth century, one of its buildings was converted to form the concert hall for the Aldeburgh Festival and the original form of ventilation cowls expanded and modified as part of Arup Associates' mechanical ventilation system. Using the ground as a source of cooling continues this long line of technical development, whilst the architectural forms of the latest addition develop a sympathetic response to the context of the existing buildings which themselves sit well in the landscape (Fig. 2.19).

2.17 The Millennium Centre, Dagenham, by Penoyre and Prasad.

2.18 Montage of ground-coupling system at Snape Maltings.

2.19 View of Snape Maltings from across the River Alde.

Such recent developments in sustainable architecture provide grounds for optimism that a more sensitive approach to the environment, at both the micro-level of site and macro-level of the Earth, is being taken up, and we will see more examples of this later in the book. Part of the reason for the success of Snape Maltings as a whole is that from a distance the building seems to rise from the surrounding landscape of reed-beds, thus blurring both the site definition and the building's outline, creating a sense of spatial and biological continuity.

Looking at any site in such a rounded manner suggests one way a sustainable architecture could evolve, by responding to aspects of more passive traditions and place at the same time as actively integrating new technologies to generate energy. A building's siting and form should take account of the sun's path to maximise not only passive solar gain but also its potential for generating energy, an approach that should also embrace the wind, rain and the ground. The overall aim of siting a building must be to develop a harmonious relationship between it and the surrounding environment, not only the immediate context but also the biosphere. This must take account of other buildings as well as natural features, topography and the way people will use and perceive the new place created. Drawing lessons from the passive past at the same time as considering the active potential of new technologies, ensures that designing the appropriate settings for the various aspects and situations of human life will not be subject to an uncritical flight to the future nor be weighed down by the past. Good design can only come from a balanced consideration of the full range of factors involved in building for people, not from a few environmental determinants. Humans need a sense of place; sites and their environmental contexts, both physical and cultural, form the basis of place making.

Building references

Millennium Centre, Dagenham: Architect – Penoyre and Prasad; Environmental Engineers – Max Fordham; Structural Engineers – Buro Happold.

Restaurant for the Aldeburgh Festival Hall: Architect – Penoyre and Prasad; Environmental Engineers – Max Fordham; Structural Engineers – Price and Myers.

3

Building design 1

Smaller buildings and the creation of environments

Introduction

Although buildings do not produce their components from their raw materials as organisms do (no building has yet made a steel girder whereas our bodies are creating bone every day), they are very similar to many living things in several ways, as we have suggested. They need to maintain a suitable environment with a stable temperature (for the comfort of their occupants) and so they take in 'nutrients' – energy, light and fresh air – and expel wastes.

We outlined in the previous chapter how buildings can also be seen as creators of environments. Through a combination of their external envelope, their structure and materials, their mechanical and electrical services systems – i.e. their physiology – they both modify the external environment and create a new internal environment which, at best, is eminently suitable for the activities of the people in the building. One of the key aspects of moving towards a sustainable architecture is to get the building itself to play a larger role, thus reducing the dependence on services – more daylight and less electrical light, for example. Another crucial aspect is to provide the energy that is required from renewable resources rather than fossil fuels.

Buildings play a vital role in the need to maintain a temperature range that is comfortable for humans, often in a climate that is too cold or too hot for comfort, as we saw with the Inuit and Matmata peoples. This is the first of two chapters on design, and deals with the house, the primary form of building. The next considers non-domestic buildings – smaller forms and larger. This division helps to clarify one of our primary themes – that buildings can be considered as similar to organisms that evolve. As in the practice of design, our overall goal is the integration of several themes including the historical, the cultural and the environmental. In terms of sustainability, the energy use of a building depends upon the overall design. Questions of daylighting, thermal capacity, ventilation and solar shading all have an impact on form, construction and appearance. This separation into two chapters allows us to develop these and other themes and also gives us room to acknowledge the differences in the types. Houses have tended to be naturally lit and ventilated, whereas in non-domestic buildings (which will be considered in the next chapter) mechanical ventilation and air-conditioning are commonplace and consequently large consumers of energy and producers of CO_2. One of the major issues for Europe in the next 30 years, however, will be the avoidance of air-conditioning in houses.

A common factor is the need to create a comfortable temperature. Humans, like all mammals, have an elaborate set of mechanisms that maintain a stable body temperature at about 37°C. These two chapters on 'Building Design' will introduce some of the factors relating to the creation of comfortable environmental conditions, for design is an act of synthesis and should be holistic in approach. More detailed aspects of environmental design will be considered in greater depth in the chapters that follow.

Plants, animals, buildings and cities all develop in a world governed by physics and chemistry. Natural selection refines living organisms and natural and human selection determine the evolution of our buildings and cities. The forms that result are of infinite variety and often of great beauty. One thing that unites them is that they function successfully (at least for a while). The trees and skyscrapers stand up and the birds and homes stay warm. The forms reflect a constant interplay between surface phenomena (for example, the need for leaves to gather light for photosynthesis or a school to let in light so that five year olds can learn to read and write) and internal phenomena linked to the volume (the need to provide oxygen to a well-protected heart at the interior of a body, for example). Physiology shows how the form and arrangement of living things is a response to their functional operations and this should be an aim of sustainable architecture. In particular it should emulate how sensitively plants and animals respond to changes that occur in their surroundings.[1]

Perhaps the best starting point in comparing biology and building is to imagine a human skeleton or alternatively a tall crane on a building site. For the most part, these are pure structures and one can clearly see what materials are used, where the parts connect, where they are subject to loads and generally how they work. Buildings under construction often show the relationship between structure and skin (Fig. 3.1).

3.1 'Skin and bone' – construction of the new Eurostar Terminal at St Pancras, London.

Now, for living animals with their internal organs and in the case of buildings with us as occupants, what one needs to do is to put a fabric or skin around the structural framework. This will provide protection and containment for the 'insides' and also provide an often complex surface which mediates between the exterior and interior, thus making a way of interacting with the environment. And the fabric and any openings in it – from eyes to mouths and from windows to doors – take on a huge functional and social importance; in the case of buildings exceptional symbolic and cultural significance also.

Animals have evolved a few basic ways of clothing their structures. Amphibians usually have smooth skins, reptiles have scales, mammals hair and birds feathers. For warm-blooded animals, the biological cost of a 'heating system' is high and so thermally efficient coverings like feathers and fur evolved. (Chapter 5 deals with this in more detail.) For buildings, the skin (which is also at times the structure) started off with a relatively small range of local materials that went from wood and animal skin to stone and earth. As technology has developed the materials involved have burgeoned to include glass, plastics, concrete, bricks, metals and timber from around the world. Each of these has its own value and significance, to which we will return.

Now another way of looking at this is to ask what happens as one goes from being a small thing that must interact with its environment (say a beach hut by the sea), or a single cell, to a much larger thing like an office building or the medieval Cathedral of Notre Dame in Paris (Fig. 3.2). The interiors of small forms are almost by definition close to their environments. And so they can survive quite well by relatively simple mechanisms for a number of their functional requirements. Insects, for example, take in oxygen at holes (spiracles) in their cuticle (outer surface) linked to air-filled tubes (trachea). At the fine ends of the trachea oxygen diffuses through to reach the blood and living cells.[2]

3.2 The Cathedral of Notre Dame, Paris.

But as organisms grow larger, while maintaining a similar shape, a number of things happen. For one, the volume enclosed increases at a greater rate than the surface area (see Chapter 4 and Appendix A) and so the structure needs to be strengthened in a similarly disproportionate way. This is why young trees look so elegant compared to older ones with their larger volumes (although mature trees have their beauty too), and why Olympic weightlifters have legs that look as if they could support the Pyramids.

Another development is that as the interior becomes more distant from the surface, it becomes more difficult to bring air into the organism or, say, light into a building. In animals elaborate systems develop to ensure an oxygen supply to every cell as, for example, our own respiratory system with complex lungs and blood supply systems. This has its counterpart in the complex mechanical and electrical systems used to provide air, heat, coolth and (artificial) light to larger buildings.

The greater volume, of course, tends to be filled. In living creatures there are more organs, muscles, blood and so on; in buildings there are more rooms, more structure, more materials and more people. This means that there is more thermal mass (see Glossary) that can be used to help regulate the temperature within the building.

From the technical point of view, a primary aim of sustainable house design in temperate Europe is to reduce the energy consumed in heating in order to lower CO_2 emissions and hence mitigate the effects of global warming. In hot countries cooling is one of the most important considerations. In countries with a variable climate there are a number of strategies. Japan, for example, experiences a short, cold winter but a longer, hotter summer. Japanese traditional architecture was therefore based on the view that people could put up with anything in the winter but summer discomfort would be intolerable. Consequently, vernacular buildings in Japan paid greatest attention to protection from the sun, typically having overhanging eaves, verandahs, open-screen windows and light, moveable screens to sub-divide interior space thereby enhancing ventilation, and making good use of planting.

An important question here is 'How do functional considerations contribute to generating the form of a building?' In the last chapter we looked at how the overall forms and arrangements of buildings responded to the particular conditions of a place, and here we will consider how exterior surfaces are developed and how interior spaces are designed in relation to the exterior. Morphologically there are a few basic ways of letting light into buildings and ventilating them without requiring much energy, and these are shown in Figure 3.3. Of course, if energy is abundant then mechanical systems can deliver air deep within buildings or, indeed, to the depths of the Earth as in deep mines.

How humans inhabit the interior needs to be considered in relation to the external environmental conditions – sun, temperature, wind – as well as views. Architecture can be considered as primarily about space and spatial relationships despite most discussion in the media – architectural and lay – focusing upon its external appearance. Clearly one aspect is that the façades must be considered in conjunction with design drawings that explore the section of a building, as an animal's skin works with its body. How a building works is best explored and described by a combination of design techniques – plans, sections, elevations, three-dimensional views and models.

However, functional considerations are only part of design. We should bear in mind that the theory of evolution developed to explain biological phenomena and that material artefacts are the product of human intention and action. Cultures can 'evolve' or

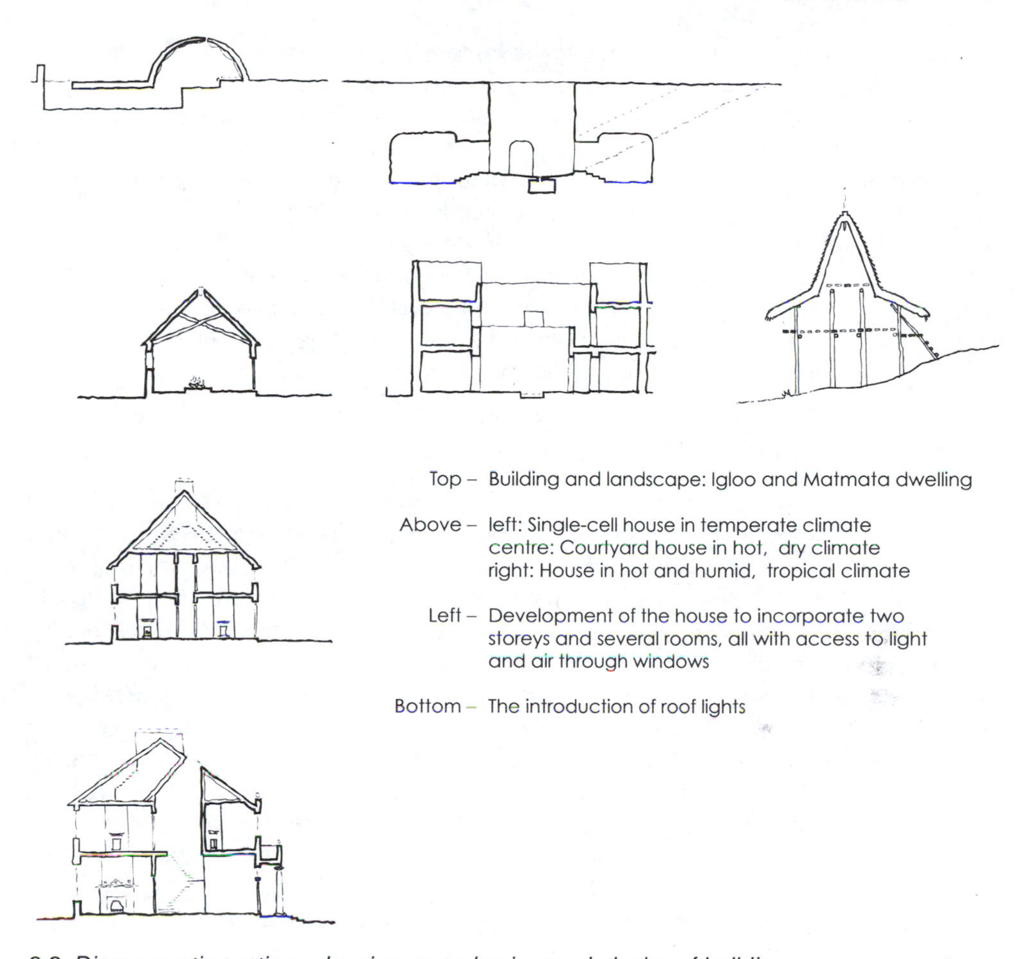

Top – Building and landscape: Igloo and Matmata dwelling

Above – left: Single-cell house in temperate climate
centre: Courtyard house in hot, dry climate
right: House in hot and humid, tropical climate

Left – Development of the house to incorporate two storeys and several rooms, all with access to light and air through windows

Bottom – The introduction of roof lights

3.3 Diagrammatic sections showing some basic morphologies of building.

develop, but can also regress or even collapse. The form of artefacts cannot be understood as being determined purely by their function, as we saw at the Neolithic settlement at Rinyo in Orkney. Ideas, the agent of cultural change, spring up in all kinds of unexpected ways. Material culture is symbolic or meaningful as well as practical. The clothes we wear, for example, give out 'messages' about what kind of person we are or wish to be. The question of meaning in materials – an important part of contemporary architectural debate – will be taken up in the next chapter.

Buildings also convey messages and design must not lose sight of the symbolic and cultural role they play. It is easy to get lost in the technicalities of producing a sustainable building and in so doing an awkward, ungainly architecture can result. A sustainable architecture should not only create comfortable conditions and use minimum energy but also nourish the human spirit – offering a glimpse of sun, a good view, the soothing effect of a breeze. Good designs must embrace ideas of beauty, appearance, appropriateness and such poetic pleasures as these for its users. Balancing technical demands with human desires means that all the dimensions of architectural design need to be explored giving due weight to the social, symbolic or cultural context.

Much post-war social housing, for example, failed in this respect, not only because of technical problems associated with newly introduced mass production but also because the environment produced looked nothing like home. Mass housing often resembles stacked boxes or packing cases. Expressing aspects of movement – i.e. lifts and stair towers – was more common than symbolising shelter or protection, as one might expect of a house. Much 1960s housing looked like neither a place of rest for its inhabitants nor buildings at ease in their environment. 'Form follows function', or 'honest expression of structure', are too narrow definitions of design for the built human environment. The new era of sustainable architecture should heed this and aim to enhance the pleasure people might have in their dwellings from sun, light and air, rather simply solving the technical dimensions of making a sustainable building.

Smaller forms: body and building

Exactly how *Homo erectus* (the builders of the Acheulian hut) gave rise to *Homo sapiens* remains clouded in uncertainty. But most paleoanthropologists agree that *Homo sapiens sapiens* – modern humans – originated in the savannah region of Africa about 100,000 years ago.[3] Environmental conditions there provided a suitable habitat for their adaptation. When *Homo sapiens sapiens* spread out from Africa to less hospitable environments, they needed clothes made from animal skins, erected windbreaks, tents and primitive huts, and found caves to provide shelter. Traces of postholes are the only evidence for the shelters of nomadic hunter-gatherers. With the discovery of farming a settled lifestyle became possible. Remains of a substantial farming settlement, one of the earliest, has been found in Anatolia, southern Turkey. Dating from about 6,500 BC, the houses at Çatal Hűyűk were built of sun-dried mud bricks.[4] Thick walls helped maintain a stable temperature through the very hot summers and cold winters of this region. Not unlike Skara Brae, these houses were single cells with no windows and only a small opening for access (rather curiously through a hole in the flat roof).

Early houses such as these and the 'beehive' houses of Greece and Ireland were essentially 'containers' – not that different from baskets and pots, or primitive organisms like crabs – into which humans crawled, sheltered and where their life took shape.[5] The word contain is defined as meaning 'to hold' and 'to keep within bounds' or 'within certain limits'. Its root is in the Latin *continere* whose stem *tenere* we saw in the last chapter means 'to hold firm' and from where the word sustainable is derived. The Latin root also gives maintain and retain, two notions fundamental to the survival of these early farmers, storing seed from autumn to spring. The first houses were essentially protection against the elements, simple containers whose thick enclosing walls helped maintain a comfortable environment by keeping out the rain and wind and moderating the temperature.

If we could fast-forward the evolution of the house we would arrive in the modern era with houses that are the antithesis of this, houses completely walled in glass. Two of the most famous examples are Mies van der Rohe's Farnsworth House and Philip Johnson's own house (Fig. 5.12). Both built in 1950's America, these could rely on cheap energy for heating. Many post-war houses pursued the Modern Movement's aim of dissolving the difference between inside and outside, seen first and most dramatically in Mies van der Rohe's Barcelona Pavilion (Fig. 3.4). For inhabited environments this could only be achieved by mechanical means of environmental control. The

3.4 The Barcelona Pavilion, by Mies van der Rohe.

suburban patio window derived from a similar aspiration. Sliding patio windows allow for ventilation but the glass wall, having a high thermal conductivity and little thermal mass, can neither prevent heat entering in summer nor retain interior heat generated in winter.

Natural ventilation and daylighting, which reduce energy consumption, have had an important impact on building form and appearance. These in turn are related to the need to control solar gains and the thermal mass of the building, factors ignored in this particular glass-walled type of extremely influential modern architecture. As we have seen, a large proportion of energy consumed is used in buildings, a problem exacerbated by this type of building. The environmental impact of a building, in particular the CO_2 emissions that it produces, depends on its overall design.

The Modern Movement was much concerned with emancipating itself from historical forms. The emphasis on abstract physical form, it could be argued, was at the expense of psychological or existential aspects of living. The first chapter of Bachelard's *The Poetics of Space* is called 'The House from Cellar to Garret; the significance of the hut'. Using images drawn from poetry, he identifies two principal themes; the appeal 'to our consciousness of verticality' and 'to our consciousness of centrality'.[6] Bachelard suggests that the vertical tiers of space in the traditional house correspond to the structure or layers of human consciousness: the subconscious, the workaday world of rational action and poetic dreams of transcendence. His book draws upon poets' memories or reveries of built spaces significant for them, spaces that conjure up images of shelter or protection. 'Great images have both a history and a prehistory', he writes, 'every great image has an unfathomable oneiric depth to which the personal past adds special colour.'[7] The spaces that have evoked such images have a special resonance for the psyche and this is an important, if elusive, aim for design today. He characterises the modern world, in contrast, as a 'bungalow age', a reductive image of dwelling to which, for all of its modernist elegance, the Farnsworth house fits.

Later chapters of Bachelard's book describe how shells and nests may have been the inspiration for human dwellings and how they have left a profound impact on the human need for spaces of intimacy and interiority. The cave and the hut recur again and again in literature expressing what appears to be a deep psychological need for shelter. Think of Thoreau building his hut in Walden woods, for example, in order to reconnect to the essential root of America. Or Mario Botta's houses of the 1970s and 1980s in mountainous Switzerland, some of which look like an entrance to a cave (Fig. 3.5). A house in Massagno (1979) has a large opening in the front wall to an internal balcony that can be closed by sliding glass doors to make an insulating barrier in winter. This chapter will briefly review the development of the house from the primitive hut to the 'machine for living in', seeking to find an appropriate balance between the technical demands of sustainability and the psychological needs of human dwelling, the advantages of tradition and advances made in modern building technology. In part this is the story of openings, the coming of windows and the elaboration of doorways. But also it will look at how traditions of planning for greater human comfort developed and were then jettisoned when mechanical systems developed.

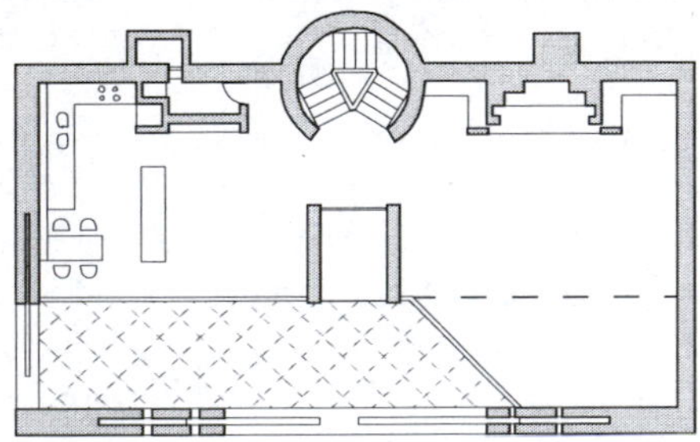

3.5 House in Massagno, Switzerland, by Mario Botta.

Culture and nature

An important role of early man's material culture was to wedge human life by symbolic means into what Mircea Eliade calls nature's 'Eternal Return' – most simply understood perhaps as the cycle of the seasons orchestrated by the sun.[8] The Dogon people of sub-Saharan Africa have baskets with a circular rim and square bottoms, for example, that represent the sun and the heavens, the symbolism of the encircling cosmos finding an echo in the house with a circular thatched roof and also the circular village enclosure[9] (Fig. 3.6). Buildings help humans understand their place in the world. In many cultures such as the Dogon it was believed that human life could only be sustained if it developed a symbiotic relationship to the natural order. This was partly practical but also had a symbolic dimension. Psychologists consider that the need for shelter stems from the human infant's need for a long period of care and protection.[10] In comparison with most animals we are born somewhat premature; the baby elephant, for example, can walk with the herd the day after it is born and the chimp, with a little help from its group, can forage as soon as weaned.

Anthropologists describe how 'pre-civilised' peoples held a world-view whereby 'nature is not sharply set off as something different from man'.[11] Robert Redfield summarises anthropological research by stating that 'primitive' 'man is *in* nature ... and we cannot speak properly of man *and* nature'.[12] Humans must have noticed, however, that articulate speech and tool use marked them out as different from other creatures in the natural world. Whilst it is acknowledged that certain apes, and possibly other creatures such as dolphins and even crows,[13] pass on behavioural traits by observation and teaching, there is a huge difference between what they do and even the simplest human cultures, which embrace reflection, art, complex tool-making and a range of artefacts. In addition we humans do not have the skin to protect us against the extreme climates which became our habitat, unlike birds and animals. Nor do we seem to have the genetic imprint to build shelter in a pre-determined way, as do birds,

3.6 Drawing of Dogon house with simplified plan.

bees or beavers. The sheer variety of houses made by *Homo sapiens sapiens* in similar climatic conditions marks us out as distinctly cultural creatures and sets humans apart from the animal kingdom where each species builds the same kind of nest more or less unchanged over time.[14]

The distinction between nature and culture has been a recurrent theme of mankind's development. We are tool users and every advance in our technology has been to gain some control over the environment. Of course mankind exploited natural resources for his own survival – a recent study of an ancient Chinese community has shown nature to have been exploited by 'rational mastery' over long periods to a point of ecological destruction.[15] Whole societies have destroyed themselves because of over-exploiting their natural resources, such as most famously on Easter Island. But more generally, 'until very recently, however, man always maintained a certain balance between his bodily and spiritual being and the external world'.[16]

Mankind's ambivalent relationship to the external world can be seen in the fact that the Chinese philosophy of Taoism stressed the search for harmony with nature rather than rational mastery. The doctrine of yin and yang explained 'the twin fundamental forces which serve to balance everything in creation'.[17] Yin became associated with clouds and rain (and the female principle), while yang became associated with fire and the heat of the sun (and the male principle). Early man 'orientated' himself with nature, not 'confronted' it. Eliade describes how something like this worldview persisted in central European peasant society in the first half of the twentieth century and, indeed, historians of English rural life recorded vestiges of this after the Second World War.[18]

The fact that humans do not have an innate capability to build shelter, as do some animals, appears as a predicament in a myth of the origin of architecture recounted by Vitruvius, the Roman architect whose *Ten Books on Architecture* (first century BC) is the only text on the subject surviving from antiquity. Amongst many things, myths are explanations of mankind's relation to nature and descriptions of the origin of cultural acts. Vitruvius writes that architecture began with the happy chance of branches fallen onto tree trunks to make a rudimentary shelter.[19] In the primitive hut we have an archetypal image of human shelter that has had a profound influence on the history of architecture.[20] Nature provided trees, happy chance (fallen branches) formed shelter, and man's reasoning powers (man, the rational animal) combined to provide the original model for building. Throughout forested early Europe – and in many other parts of the world – early buildings were not unlike that described by Vitruvius, timber framed – posts, lintels, purlins and rafters – with infill panels of thin branches covered with whatever material that was ready to hand, as was the case with the LBK houses and Neolithic houses in Britain before Skara Brae.

The Vitruvian account of the primitive hut has echoes of what Darwin wrote in his book *The Variation of Animals and Plants under Domestication*. Human intention, observation, imagination and calculation transformed the 'found' primitive hut into high architecture just as selective breeding enhanced the principle of natural selection that Darwin identified. The domestication of species has produced a wide range of plants and animals that suit human needs and tastes – in the case of dogs, for example, from dachshunds to Great Danes. In a similar way many cultures have produced a wide range of buildings to suit their particular tastes as well as adapting to their climate and available materials.

Two evocative examples of how different environments helped shape distinct building forms are the Swiss or Norwegian log cabin and the English half-timbered house.

3.7 Half-timbered English farmhouse (left) and Norwegian log cabin (right).

The characteristic low-pitched roof of the Swiss and Norwegian house lets snow settle on it in winter, which provides some insulation against the cold. Its simple roof form is to prevent melting snow from seeping into the building as it would if there were valley gutters, and its wide overhanging eaves ensure that water dripping from the roof does not splash back and rot the softwood logs of the wall (Fig. 3.7). The English half-timbered house, in contrast, is made of hardwood and relatively thin timber sections to make a frame. The walls were often in-filled with wattle and daub, a woven panel of thin branches covered with clay or plaster. Its characteristic steep-pitched roof with gables reflects the prevailing rainy climate. The half-timbered house looks at home in its English woodland context as does the Swiss log cabin on the forested slopes of the Alps. Both these very different examples demonstrate the general principle that good architecture involves the integration of building and the broader environmental conditions of the site.

Alvar Aalto said that the indigenous buildings – log cabins, similar to those in Norway – of the thickly forested region called Karelia paralleled the collection of Finnish founding myths collected in the *Kalevala*. Left to itself, he wrote, this region produced buildings whose 'construction has therefore been dictated directly by natural conditions'.[21]

> Karelian architecture (shows how) human life can be harmoniously reconciled with nature at our latitudes ... It is a forest architecture pure and simple, with wood dominating almost one hundred percent both as a building material and in jointing. From the roof, with its strong log structures, to the moving parts of the building, we find timber, which is generally left naked, without the effect of immateriality given by colouring. Timber is generally used as close as possible in its natural size, according to its own scale.[22]

The sturdy trunks of pine provided material for the building and branches were used for making furniture, exploiting bends and knots, 'parts ready made by nature'. The way Aalto's particular form of regional modernism developed was derived from his distinctly biological conception of traditional building, which we will see more of a little later on.

> The Karelian house is a building that begins with a single small cell, or dispersed embryonic shacks – shelters for people and animals – and grows, figuratively

> speaking, year by year.... It is comparable to a biological cluster of cells; the possibility of a larger and more complete structure always remains open.[23]

Only in the past 300 years or so since the Enlightenment has mankind operated by a belief in his separateness from nature. This attitude, which some see as beginning with the philosophy of Descartes, has come to dominate Western thinking and its economic application has spread all over the globe. Descartes' method of doubt led to his famous phrase 'Cogito ergo sum': I think therefore I am. From this he came to consider that the mind was quite distinct from the body. 'By locating the point of certainty in my own awareness of my own self, Descartes gives a first person twist to the theory of knowledge that dominated the following centuries.'[24] Cartesian rationalism accepted as certain knowledge only that which convinced by pure reason, logic and mathematics (including geometry) or through verifiable experiments. The consequence of what became known as the 'mind–body dualism' is that the world is represented as an object of thought. With mind the determinant of truth it was but a short step for man to see himself as the hub of reality and the Earth as a mere stockpile of resources.[25] In many earlier societies, some aspects of material culture aimed to express a symbiotic relationship with nature. This is something a great number of people today feel in their private life and leisure activity. The ecological crisis is causing us to rethink our attitude to nature in our economic and working practices. Our power to take from nature is now immensely greater than when we had no more than wind, water and animals. With power must go responsibility. A sustainable architecture should not only be ecologically sound in practice but also express a more sensitive relationship between human activity and nature.

The single-cell house evolves to a complex organism

As houses developed, similar forces to that determining the size and shape of animals were at work. Despite being as different in appearance as the camel and the polar bear, most vernacular houses were constrained to be no more than two or three rooms deep to admit light and air, and two or three storeys tall because of structure. Just as large animals evolved internal organs to distribute essential nutrients, so the expansion of the house by the addition of more cells was more or less limited by the requirement to get air and light into each room. And as we will see later, cities had to evolve in similar ways to let in light, air and materials and get wastes out.

As societies developed, traditions of houses arose to accommodate the more complex arrangements that people desired and to provide greater comfort. We can trace this through vernacular building where, taking the English house as an example, we see the medieval hall house adding rooms for specific functions to the earlier general-purpose hall. Of its ancestor, the Saxon house, only a few post-holes have survived, but similar houses exist in Friesland in northwest Germany, where the Anglo Saxons originated. In this type of dwelling, all the necessities of settlement are under one roof, house and barn combined. Humans lived in one corner, calves in another, cows along one edge, and hay filled the centre. Animals brought in for the winter added heat to the dwelling, hay possibly provided some residual heat, and may have been used to insulate the living quarters. (Straw bales have recently become used as wall material for some radical 'green' houses. See Appendix A.) Saxon buildings were timber framed with walls built of split trunks of oak fixed vertically often on a plinth of

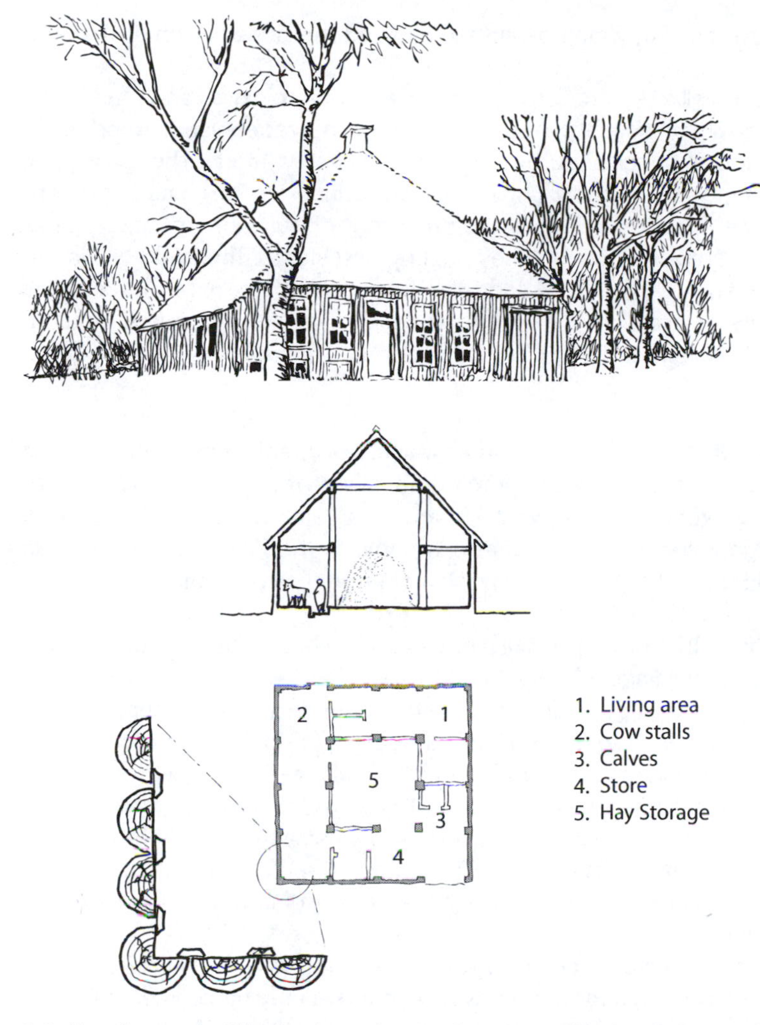

3.8 Farmhouse in Friesland, northwest Germany. This is similar to Saxon houses in ninth-century Britain. The detail is of a Saxon log wall church at Greensted, Essex.

brick or stone (Fig. 3.8). Farms were clearings in a forested landscape and wood was all-important, providing not only building material but also fuel for fire and acorns for pigs. The significance of wood can be seen from the fact that the Saxon noun for a building was *timbrian* – literally meaning timber – and the verb to build was *getimbrian*.[26]

The great hall of Stokesay Castle (thirteenth century) illustrates the cavernous nature of the central living space of a medieval house with its open timbered roof (Fig. 3.9). Like Saxon and earlier houses, Stokesay retained the original open fire that burned at the centre of the hall with a smoke hole in the roof – timber-framed farm-houses with this arrangement can be seen at the Weald and Downland Museum near Chichester. Gradually fireplaces and chimneys were introduced into what must have been very cold and draughty spaces. The fireplace provided heat to where it was required and the brick fireback and chimney stored heat. The primary function of the chimney was to remove smoke, which eventually allowed the upper volume of space to be occupied.

Fourteenth-century dwellings such as Haddon Hall in Derbyshire and Penshurst Place in Kent show how the house evolved into something recognisably modern in terms of rooms for more specific purposes and greater comfort. The great hall remains open to the timber roof but a timber screen is inserted to make a draught lobby from the entrance (Fig. 3.10). This is the precursor of our modern, small, more purely functional hall. A fireplace and chimney occupy one side of the great hall in the medieval house. Even so, the houses would have been draughty. Older readers will remember the difficulty of avoiding cold draughts in houses before central heating. C.P. Snow's novel *The Masters* evokes something of the atmosphere of a medieval hall in winter.

> It was scorching hot in front of the fire, and warm, cosy, shielded, in the zone of the two armchairs and the sofa which formed an island of comfort round the fire-place. Outside that zone, as one went towards the walls of the lofty medieval room, the draughts were bitter ... so that one came to treat most of the room as the open air, and hurried back to the cosy island in front of the fireplace.[27]

(Recent revisions to the UK Building Regulations have made a high level of air tightness mandatory in new buildings. Air tightness in building design is primarily concerned with eliminating air leakage, thus reducing the amount of energy consumed in heating. Bill Dunster's and Bio-Regional's BedZed housing is designed for high levels of air tightness which, combined with high insulation levels, reduces heat loss to the point where a conventional heating system is not required.[28] Several other recent houses have been built without heating by using 'superinsulation', air tightness and orientation towards the south to maximise solar gain.[29] Greater understanding and control of air movement is crucial to retaining heat and providing comfort and we will return to this in Chapters 5 and 6.)

Attached to one end of the medieval house was a buttery, pantry and kitchen, and to the other a parlour – a small room for more private conversations, sometimes called by names such as chamber or *bour*. The relative intimacy of this space finds an echo in the word bower, now applied to a cosy shelter in a garden, a picturesque image of the primitive hut or original shelter. Above the parlour is a *solar*, which was the bed-sitting room of the proprietor. The name indicates a direct relationship to the sun, detectable in the relatively large south-facing windows. The name *solar* was originally applied to an upper floor having an open roof where ancient Greeks and Romans sunbathed.

The honorific role it played combined to keep the lofty, open timber-roofed hall a major feature of houses until the fifteenth and sixteenth centuries. However the advantages of ceilings became apparent – sealing the lofty roof space reduced the volume of air and consequently heat lost to the upper, unoccupied part. Not only did inserting a first floor over the hall provide greater comfort but also more useable space. The great open-roofed hall became replaced by single-height rooms with wood-panel lined walls and plaster ceilings in grand houses of the sixteenth and seventeenth centuries. Massive oak beams and posts supported floor joists to create an upper storey in the more pragmatic farmhouses and inns of the period. This picturesque arrangement is found surviving in many country pubs, as are settles, high-backed wooden furniture designed to shield people from draughts.

3.9 Stokesay Castle, Shropshire.

1 – Hall
2 – Line of gallery
3 – Dais
4 – Parlour
5 – Stairs up to solar over parlour
6 – Buttery
7 – Corridor to kitchens
8 – Wine cellar
9 – Entrance porch
10 – Upper courtyard

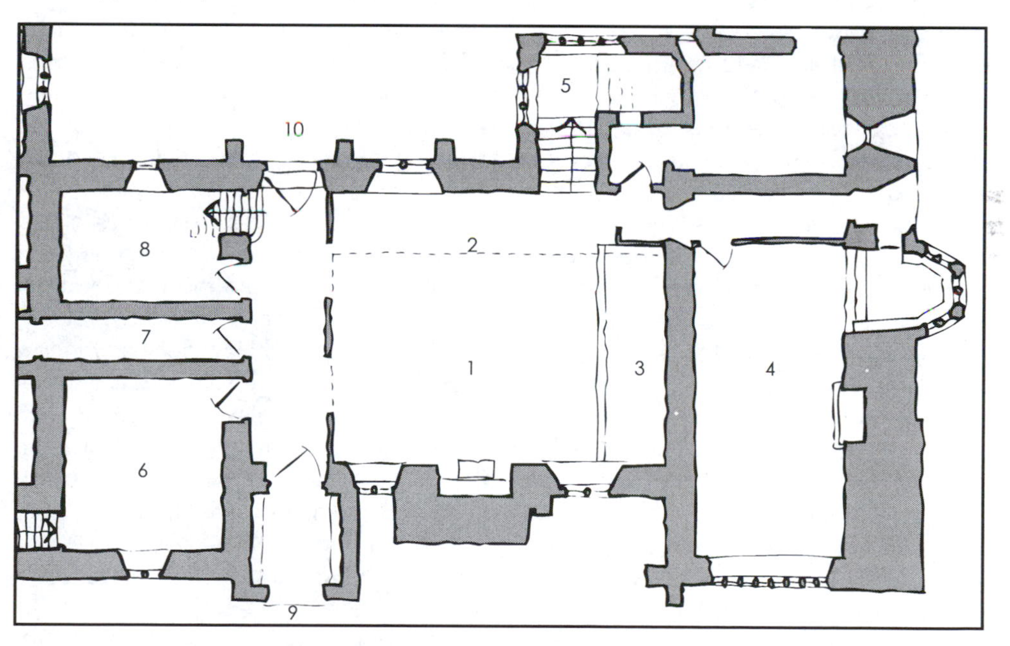

3.10 Ground-floor plan of Haddon Hall, Derbyshire.

The origin of the word window gives insights into its introduction, use and development. Medieval English *windoge* is derived from the Old Norse *vindauge*, which means 'an eye of the wind'. From this it is clear that windows were connected with both light and ventilation, although the latter may have been less desirable then in the draughty Saxon halls than now. The first windows would have been simple holes in the wall. Later English examples of open windows, with vertical oak bars set diagonally in the opening, can be seen in smaller fourteenth- and fifteenth-century farmhouses restored at the Weald and Downland Museum. Curtains, hangings or internal wooden shutters gave some protection against wind and rain in these earliest windows. (In southern France sacks filled with straw were sometimes stuffed into window openings during the cold season.)

In his *History of the English House* Nathaniel Lloyd writes that there are no records of glass in British houses before the twelfth century.[30] Frequent references to glass occur shortly after this indicating its novelty and increasing availability. The first glazed windows comprised many small panes of glass held together by lead strips, or 'cames'. Just as the single volume of the hall-house became sub-divided for comfort, so did the window. Stone or timber mullions were inserted into the window opening and a side-hung openable casement was fitted into one or more of the lower sections, which provided ventilation. By the end of the seventeenth century, casements began to be superseded by sash windows. The vertical sliding sash proved to be popular because the amount of ventilation could be finely controlled, giving small or large openings at the top or bottom of the window, or both top and bottom. These relatively fine-tuned arrangements helped clear smoke that rose to the ceiling.

Associated with vertical sliding sash windows were shutters that folded back into the window reveals. These not only gave privacy but also had some insulating value. With the new light and rigid insulating materials available today, there is scope for inventive reworking of this particular feature of eighteenth-century windows. In hotter parts of Europe shutters were traditionally placed on the outside of buildings and used to prevent excessive solar gain. They also gave some privacy where dwellings faced one another at close quarters across the narrow streets favoured in hot climates.

Windows came to have more than a purely functional role in providing light and ventilation, particularly during the Georgian period. The Georgian town house is characterised by a carefully designed set of proportions that modulate the façade into a harmonious balance of solid and void. Derived from the proportional systems of classical architecture, the way to provide 'correct' related dimensions for a door or window was made accessible to the builders of Georgian houses through pocket-size handbooks produced by such as the wonderfully named Batty Langley and William Halfpenny.[31] Proportions laid down in these late-eighteenth-century handbooks centred on the double square, although doors were often more than twice as high as wide.[32] Georgian doors usually have a fanlight above to help light the hall. The proportions are generally vertical reflecting the upright stance of the human body (Fig. 3.11). The word façade is clearly related to face, and windows and doors play an important role in establishing the character of a building, just as do eyes and mouth to a face.

The size of doors and windows add to the sense of human scale in buildings, for they relate to human use and express the quality of interaction between inside and outside. Not only was the size of a building constrained by these needs for light, air and access, but also by the length of timber available for roofs and floors as well as the capability of men to lift them into place. In these aspects of traditional and vernac-

3.11 A Georgian house in Soho Square, London.

ular building probably lie the origin and deep significance of scale in buildings. The invention of the arch, of course, modified this, but it is only since the Industrial Revolution, the production of steel and sophisticated methods of calculation that we have broken free from an order determined largely by the measure of man.

Making buildings that respond sensitively to human scale should be an important part of a sustainable architecture, helping to focus the relationship between humans and the planet. In most countries the roof expresses the protective role of buildings in vernacular traditions, but the walls also contribute a great deal. The proportion of wall to windows, solid to void, plays a large part in communicating the feeling of whether a building is open or closed, more protective or less. Openings in the wall have an important expressive or symbolic role to play and this aspect should be kept in mind as the more quantifiable problems of designing internal environments are broached. A poetic or fulfilling sustainable architecture should aim to be a symbolic mediator between mankind and the broader environment.

A loose analogy might also be made between bodies and buildings in comparing the lung with the courtyard. The lung supplies oxygen to the body as light and air are supplied from the courtyard to the interior of a building. The Roman house had rooms and colonnade surrounding an open court, although this may have been covered in northern climates such as Britain. (It was once thought that the medieval central hall had derived from the Roman house with its central court or atrium, but there is no evidence of Saxons taking over Roman dwellings.[33]) The courtyard was derived from Greek and Asiatic prototypes that evolved to cope with hot climates. The role of the court was to provide a space shaded from the sun during the day and where cool night air could settle and linger (Fig. 3.12). Plants and water were often used to enhance the cooling effect through the processes of transpiration and evaporation. The house surrounding a small courtyard, often with plants, fountain or a pool of water, remained a traditional type for hot climates such as found in southern Spain or North Africa. (The close connection between plants and man is perhaps epitomised by the Japanese term for home, which is a combination of house plus garden.[34]) Modern examples of courtyard apartment buildings in Seville use canvas blinds on straining wires to provide controllable shade (Fig. 3.13). We will return to the courtyard in the next chapter when we look at the evolution of larger forms of building.

Vernacular forms of construction became somewhat sidelined by the emergence of a Classical tradition during the seventeenth and eighteenth centuries and the pursuit of style revivalism in the nineteenth. But at its end the rise of the Arts and Crafts movement saw a new interest in the vernacular. The country house raised this to its zenith, combining traditional forms of construction with sophisticated planning of rooms for a wide range of functions and specific relationships to the exterior. The Arts and Crafts house can be seen as a late outcome of the Romantic Movement. Philip Webb described the vernacular as 'folk-art ballad building', probably a reference to Wordsworth's and Coleridge's *Lyrical Ballads*, a seminal work of Romanticism.[35] The Romantic Movement emphasised human engagement with the natural world, hence landscape, the picturesque and mood were major concerns. A certain flamboyance and excess often characterised Romantic work (although not Webb's), in contrast to the severity of approach of the Classical spirit, which emphasised ideas about refinement, purity and paring down. The Classical spirit played a part in the emergence of Modernism, with its emphasis on whiteness and the framed building, after the romance of the Arts and Crafts.

Architecture can suffer from a slavish adherence to ideologies, or a narrow, purely aesthetic belief in a style. Some Arts and Crafts architects, such as Edward Prior (whose Home Place we discussed in the previous chapter) and William Lethaby, were both Romantics and rationalists, combining the newly rediscovered material concrete with traditional, craft ways of making buildings. A sustainable approach would benefit from considering both the Classical and Romantic spirit, just as we believe it should draw upon the vernacular and modern technology.

Lethaby's Melsetter House (1896) in Orkney, one of the last great country houses, demonstrates how a sophisticated tradition of planning for greater comfort evolved (Fig. 3.14). Lethaby believed that architecture should be the development of types

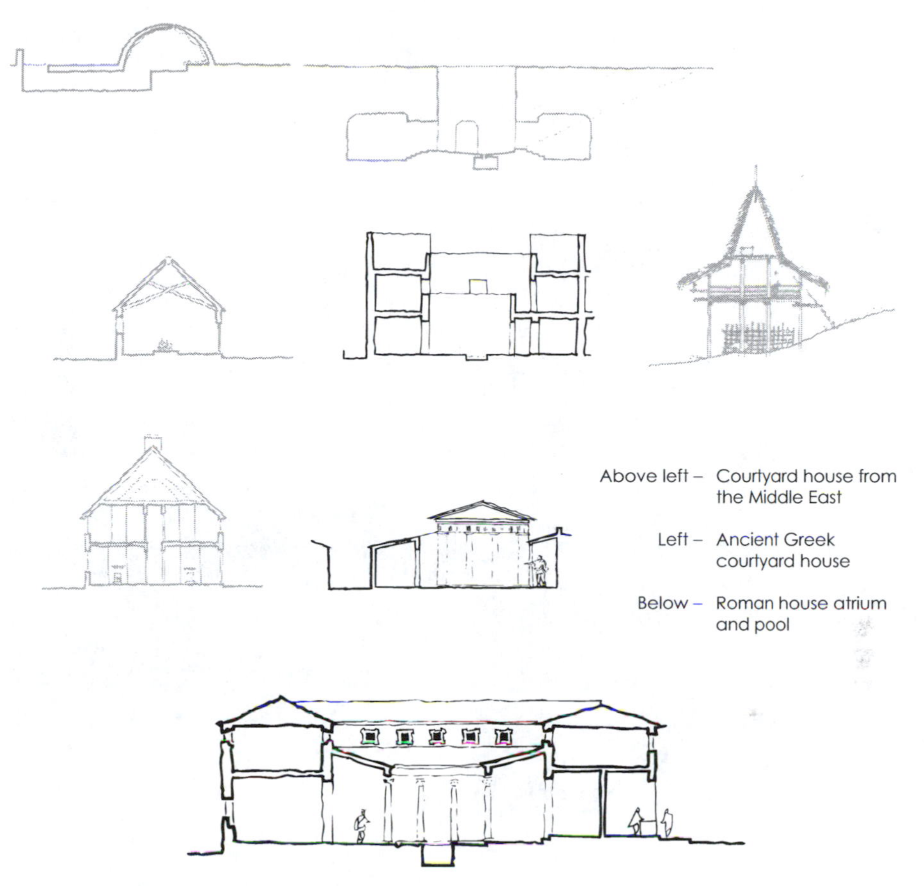

Above left – Courtyard house from the Middle East

Left – Ancient Greek courtyard house

Below – Roman house atrium and pool

3.12 Diagrammatic sections showing the development of courtyard-type dwellings.

3.13 Courtyard housing in Hombre de Piedra St, Seville, by Cruz and Ortiz.

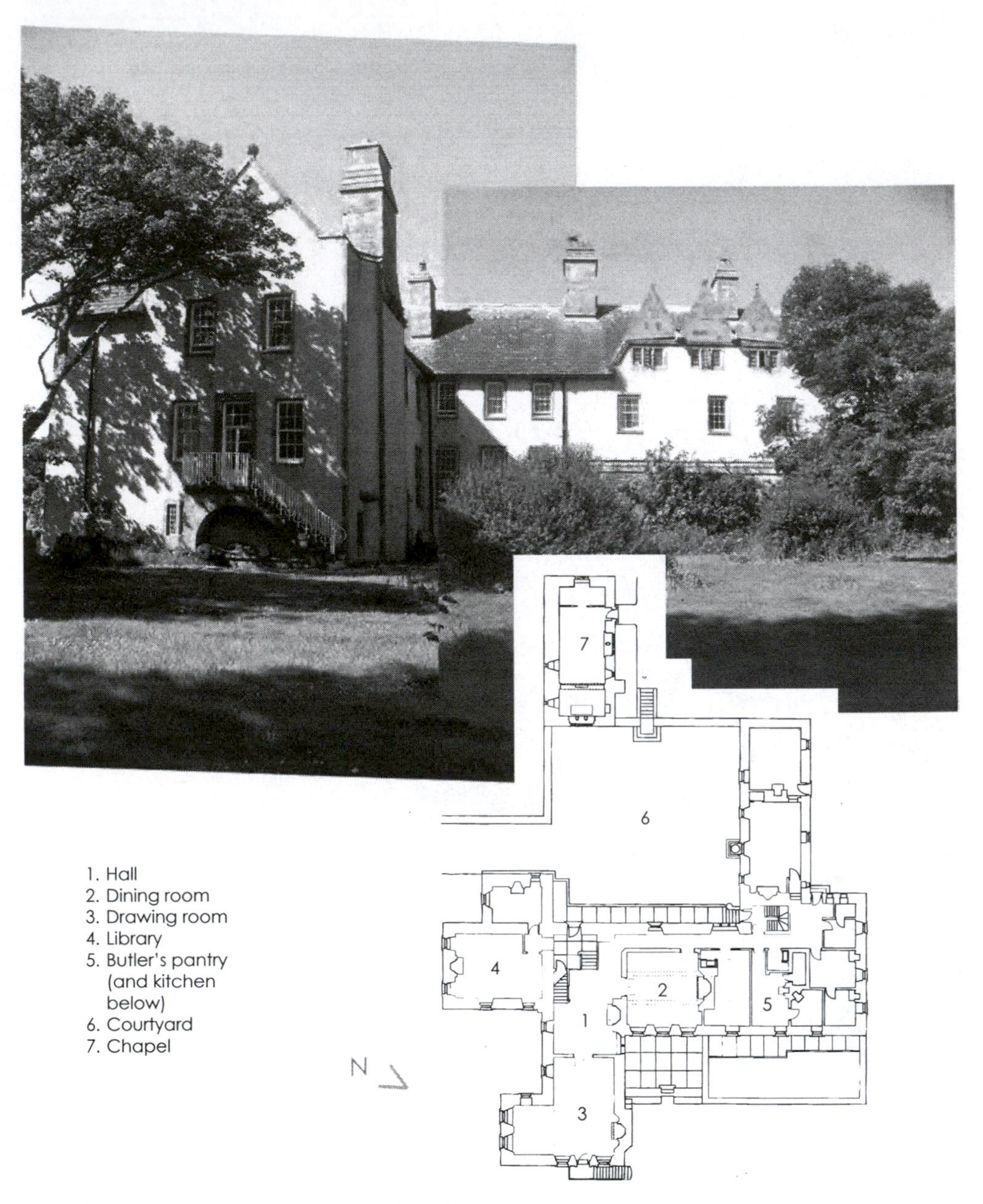

3.14 Melsetter House, Hoy, Orkney, by William Lethaby.

ready to hand. Melsetter House continued a line of development whose plan arrangement was crystallised by Webb in the Red House (1859) for William Morris, which became the prototype for many of the houses built by Arts and Crafts architects.[36]

In essence, the planning of a building might be thought of as inviting in what is desirable – some sun, breezes, views – and excluding what is undesirable – cold winds, rain and excessive sun. This traditional role is now supplemented by the need to 'invite' the sun onto its surfaces to provide solar energy. In his book *Das Englische*

Haus (1904), Hermann Muthesius explained to his German readers how the organisation and form of Arts and Crafts houses were a perfect response to the Victorian way of life, being well adapted to the country and the climate.[37] The planning of Melsetter is a useful illustration of Muthesius' argument, despite Lethaby modifying and extending an existing house. In addition it illustrates how more recent buildings than Skara Brae responded to the environmental conditions of Orkney. The thick-walled house is sited beneath a hill that shelters it from the prevailing west winds and is gathered around a courtyard to provide further protection. Protection from the wind was always a major concern in tree-less Orkney as we saw at Skara Brae. (Today wind has become a potential source of energy with a wind-farm built on top of one of the highest hills. Less contentiously, and more relevant here, a house and visitors' centre on North Ronaldsay – the northernmost Orkney Island – has been heated by electricity generated from two small wind turbines. Issues of scale and the appropriate siting of wind turbines will be considered in Chapter 8.) The stone walls of Melsetter are covered with harling – a roughcast render – to provide a more waterproof skin.

Muthesius says that the sun should enter the dining room only at breakfast and – as the plan with sun path diagram superimposed shows (Fig. 7.5) – only those who took an early breakfast at Melsetter would obtain a glimpse. The drawing room (where much of Edwardian day-to-day living took place), in contrast, has more windows and is open to almost all of the day. It projects out into the sun's path and has a small, high level window in its west wall to catch the early evening sun (Fig. 3.15). These two rooms enjoy the best views over the garden and Scapa Flow. This group of rooms organised around particular views and orientation well illustrates the art of good planning, which involves making complex and balanced choices rather than being determined by a singular cause. Sash windows were the preferred choice of Arts and Crafts architects not only because they give a wide range of ventilation with either high or low openings, but also they seemed modern compared with Gothic Revival windows much favoured at this time.

The library projects out to the south so that the sun's warmth dries the book-lined walls throughout the day. The hall is a relatively internalised room in country houses and often alluded to its medieval ancestor. At Melsetter the stair is planned in such a way as to open up a double height space down which the evening sun can pour in summer. Long summer vacations were a feature of late-nineteenth-century country houses and the hall, a formal statement of the owner's aspirations, was a place where guests gathered before dinner. In his novel *The Other House* (1893), Henry James remarked how the gentlemen would admire the dresses of the ladies as they descended the stairs.

Each of the rooms is centred on a fireplace, particularly impressive in the hall. Striking the eye on entering, the dominating sandstone chimneypiece adds to the theatricality of everyday life that James described. The dining room has a shallow inglenook for gathering close to the fire, a tradition elaborated to impressive effect in the houses of Baillie Scott in particular. The inglenook came to symbolise the idea of shelter in these Arts and Crafts houses. In the Southern States of America at about this time, an Arts and Crafts tradition of houses developed that were surrounded by verandahs, which allowed people to live part of their lives on the shaded and airy cool edge of the building.[38] The idea of a house that allows for deep retreat in winter and opens up in summer is one that is worth revisiting and one made easier by modern technology.

The bedrooms at Melsetter are arranged in the plan to receive the early morning sun. Beds were placed in such a position as to avoid draughts between door and

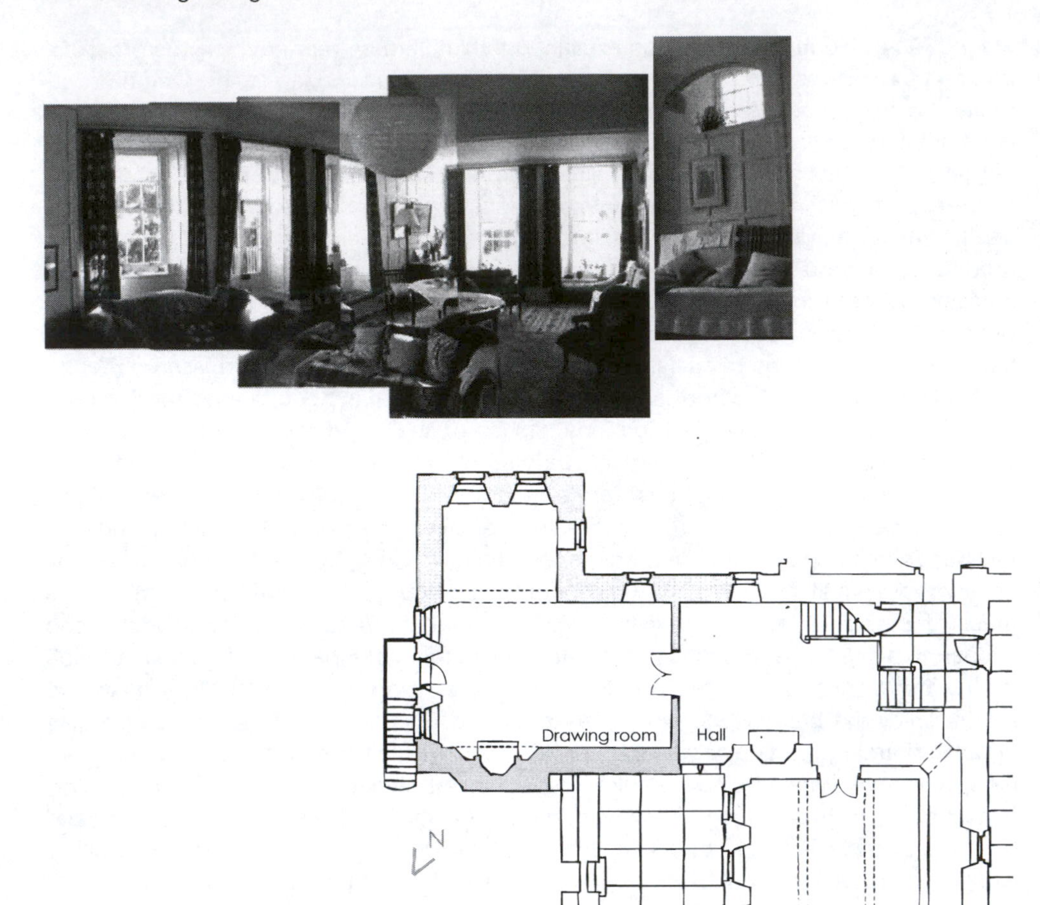

3.15 Drawing room at Melsetter House designed for sunlight throughout the day.

fireplace or door and window. Because this was the era before refrigerators, the kitchen and other service spaces for storage and preparation of food were located in the cool north or north-east corner of these houses.

Our own tastes or ideas on living today are different, of course, from that described by Muthesius. For example, we might prefer to have the sun where we dine in the evening when we come home from work. But the basic principle of planning a building by careful consideration of particular activities and shaping space appropriately around them remains relevant. Inviting in what we desire and excluding what is undesirable through judicious planning should form part of a sustainable approach to architecture and planning as well as the more recent concerns with reducing energy, etc. These two aims go together, of course.

The Prairie houses of Frank Lloyd Wright began from the 'Shingle' or 'Stick' style (as the Arts and Crafts tradition of American houses is often called) but evolved looser plans in a similar way to Melsetter. Wright's houses are considered as pioneers of Modernism because distinct rooms give way to free-flowing space and deep overhanging eaves, suggesting a blurring of inside and outside, or a distinct transitional zone between. Although a massive chimney still occupies the centre of his houses, this was more symbolic than practical. The idea of open-plan dwelling was made possible by Wright's use of the recently developed central heating systems, in which a central boiler and pump deliver heat to radiators distributed around the house.

The overall spreadeagled shape of Melsetter House plan and others of this period show their form being shaped by the search for appropriate light. Because the deep root of the large country house was the 'smaller form' of the vernacular farmhouse, the complex plan shows its struggle to evolve and adapt to a new, more modern primary aim of pleasure in light and views. It is a thick-skinned creature, its harling, stone walls, its gables and the relatively small proportion of window to wall, express its fundamentally defensive character against the cool, wet climate. Most British houses were cast in this mould until the advent of central heating and double-glazing after the Second World War. These technical possibilities coincided with more leisure and a widespread desire to engage more with the sun and air. Holidays were for this and the home followed suit. Recent changes in the UK Building Regulations are tending to restrict the size of windows with the aim of reducing heat loss.

Muthesius' book conveyed Arts and Crafts ideas on planning and technical advances in building to Germany and the continent where they played a part in the development of the Modern Movement. But the rational planning for comfort and creating appropriate settings for different situations that characterised the English Arts and Crafts became supplanted by the idea of rationalising production, a particularly important aim of the Bauhaus (1919–33). Houses such as Melsetter represent the culmination of a long tradition whereby the organisation of the building itself, the location of rooms and windows carefully orientated to the sun, produced comfortable living conditions in response to particular climatic conditions. But late-nineteenth-century technical developments, in conjunction with the increased part played by non-domestic buildings, changed this (as we will see in the next chapter). These houses also became unsustainable from a social point of view. With every room needing a fire, for example, the household needed many servants to keep it going. The country house evolved, or fossilised, to become a National Trust monument. And domestic machines powered by the 'genie' of electricity were invented to do the work of domestic servants.

House and machine

Early in the twentieth century Le Corbusier described the house as 'a machine for living in'. His catchy phrase found unlikely support from Lethaby, although he stressed the 'living in' rather than the machine, in contrast to Le Corbusier and his followers. Le Corbusier's Maison Cook (1926) drew its imagery from the Farnsworth Goliath biplane, his Villa Stein (1926–27) looks like a ship's bridge, his Villa Savoye like an unearthly flying machine (Fig. 3.16). As we have suggested, however, a house is not simply a machine, it is not simply an efficient organisation of spaces and the sum total of domestic machines, but something deeper, less easily grasped but equally important. From his wide reading in anthropology, Eliade says that 'the house is not an object, "a machine to live in", it is the universe that man constructs for himself'.[39] The house is central to the way humans have developed a relationship to the world.

Inspired by Le Corbusier, Aalto designed the Paimio Sanitorium (1928–33) in the streamlined International Style. Subsequently he reconsidered this position to design buildings that remained recognisably modern yet were informed by the vernacular traditions we discussed earlier and that were adapted to the particular characteristics of the Finnish landscape. His Villa Mairea (1937–38) (Fig. 3.17) can be read as an overt

3.16 Villa Savoye, Poissy, near Paris, by Le Corbusier.

3.17 Porch as aeroplane wing at the Villa Stein, by Le Corbusier (right) compared with Alvar Aalto's at the Villa Mairea.

adaptation or critique of Le Corbusier's Villa Stein. Significant additions express the way that he considered the cultural and environmental roots required of buildings that satisfy human dwelling, in the deepest sense. Although perhaps not a sustainable building in the modern, technical sense, it remains a fine example of how a building can respond to the particularities of a place and mediate between human life and landscape.

Where Le Corbusier announces entry with a canopy that looks like an aeroplane wing, Aalto designed a porch supported on unsawn timber poles tied together with rope and supported on boulders, an entrance that links the house with the forest in which it sits. Immediately on entering, a balustrade with an irregular rhythm of full-height balusters evokes the random pattern of trees in the forest. Steel columns are wrapped with rattan, which looks like bark peeling from a birch tree. The floors are wood and a moveable plywood screen separates the living area from the study. (Aalto imaginatively exploited the technical advantages of plywood, steam bending and laminated timber in his furniture design as well as in buildings.) The living room looks over the garden to a grass-roofed sauna, like a primitive hut in the forest clearing. Inspired by the arrangement of Finnish farmsteads, Aalto designed the Villa Mairea as an L-shaped plan that helps engage the house with its garden. The overall appearance of the house clearly has its origin in International Style whiteness, but there is an overlay of slate and ply in places, and individual windows with wooden surrounds, which are oriented obliquely to the best light, replace the canonical strip window preferred by Le Corbusier. These details, and more, define the kind of architecture that allows humans to grasp something of their relationship to the world about them, promoting a better understanding of how our existence is bound up with planet Earth in the deeper sense of dwelling implied by Eliade's statement above.

Two legacies of the Modern Movement have been the machine-like vessel independent of its environment and the building as a glass box. A phrase associated with Mies van der Rohe and this phase of Modernism was 'Less is more'. Given the clutter of nineteenth-century houses and ubiquitous ornamentation, this was an understandable reaction. The actual experience of living, however, with all its rich complexity and range of different spaces that have evolved in response to different needs, argues against this as a definitive statement. Doing less with more, using less energy than more, making spaces that enrich human experience by the use of materials, textures and good natural light and the support of a few beautiful objects; these, however, are propositions to which a sustainable architecture might subscribe.

Architecture has evolved a rich language of forms and details that combine to allow a building to be 'read' as an expression of how human societies, at different times and in different places, have engaged with – and sheltered from – the world about them. When this is reduced to a simple, unarticulated glass wall or a sealed skin, it not only creates problems of solar gain and heat loss but also eliminates all the subtle nuances that constitute a refined language. Façades are an essential part of the building aesthetic and operation. As well as being the visible face of the building to the outside, the façade is the boundary and threshold between the internal occupied spaces and the external environment.

The vernacular house turned its back on the outside to shelter its inhabitants; the Modern Movement opened up to light and space. Sustainable architecture should aim to be a subtle interweaving of both traditions. In addition, design will increasingly involve renewable energy sources, which we will consider in the next chapter.

Buildings should be designed to suit the climatic conditions of their particular place. Traditional or vernacular buildings did this, as we have seen, and now more than ever architecture needs to respond sensitively to climate and imaginatively to the possibilities offered by new materials and energy generation.

The Villa Mairea and the work of Glenn Murcutt are good and varied examples of how this can be done in part. The development of Murcutt's work shows a dramatic reversal of the energy-consuming glass box. His first houses – such as the Lawrie Short House (1972–73) – were inspired by Mies van der Rohe – and made tenable in the hot Australian climate by air-conditioning – whereas his more recent designs evoke ideas of shelter and protection, as we saw in the last chapter.[40] Although these are still made of metal and glass – the quintessential materials of the modern age –

3.18 Ball-Eastaway House, Glenorie (1980–83), Australia, by Glenn Murcutt.

they return us to the evocative form of the primitive hut. The house has walls of glass louvres and timber screens that allow breezes to blow through, a double-skin roof with rotating ventilators to help remove the heat, and rainwater pipes and gutters are given pronounced expression (Fig. 3.18). In this way he builds up a poetic language that resonates with archetypal ideas of shelter as well as responding to a particular place. His designs are good illustrations of Frampton's 'Critical Regionalism', a cross-fertilisation between a rooted vernacular and aspects of the modern 'universal civilisation' that inevitably impinges in an age of rapid change and global communication. Extending this notion a little further, the morphology and materials of traditional buildings from a particular region can serve as a good basis for integrating the scientific advances in environmental design. Murcutt's buildings are rich in cultural references but are also economical with material, as is nature, and we should be also.

4

Building design 2

The environments of larger buildings

Introduction

The Arts and Crafts house marked the culmination of a tradition of good planning that was broken by the onset of the Modern Movement. As we saw in the last chapter, human habitations evolved from simple, single-cell structures to the complex organism that was the late-nineteenth-century country house. The draughty medieval hall-house gradually gave way to a more comfortable arrangement of rooms shaped for specific roles and orientated to the external environment for optimum conditions of comfort and views. The vernacular root of this development made economical use of local materials for building as well as sources of energy. Energy remained scarce for most people in the nineteenth century, but was abundant for the few who could afford it (such as country house owners with servants to tend the many coal fires). A rapidly changing society demanded new building types which combined with an abundance of fossil fuel, developing technologies, changing tastes and fresh ideas over the years to generate buildings less responsive to the environment. The Modern Movement arose in this era of cheap energy following the discovery of oil and its widespread use in powering the building services in the early twentieth century.

The character of much Modern Movement architecture was shaped by the rise of Positivism – the belief that everything should be subject to scientific forms of analysis – at the end of the nineteenth century. This approach when applied to planning emphasised the new science of ergonomics and the efficient relationships of the internal arrangements of a building. Another ambition of the Modern Movement was to dissolve the sense of enclosure in building. Space would flow like a fluid medium between inside and outside. The glass wall made this appear possible and, aided by technical advances in heating and cooling in conjunction with inexpensive fossil fuels, a comfortable temperature could be created in any climate. Prior to this, as we have seen, new building types had developed by processes of adaptation not unlike the slow interactive processes of biological organisms. The Modern Movement rejected historical expression and the avant-garde promoted entirely new forms of building.

Our own time is one of dwindling fossil fuels and climate change. These two facts force us to consider how to design buildings that not only use less energy but can also generate part of their energy demands. Modernism's continuing legacy of making comfortable environments using the power of technology – mechanical heating, ventilation and air-conditioning – is one that needs to be reconsidered in a sustainable architecture which is more attentive to its impact on the biosphere.

This chapter will outline the history of these developments. It will also consider the question of materials in sustainable design where there is now immense choice. Industrialisation produced a huge growth in building and cheap transport made the selection of material primarily governed by its cost rather than what was locally available. As we have seen, cheap Welsh slates and other materials were transported all over Britain and replaced local or regional materials. The broader environmental impact of burning coal and rapid transport was unknown at that time. The complex issues of material choice will also be considered in the light of what the qualities of materials might mean to the human experience of the built environment. The chapter will conclude with some thoughts on the evolution of future forms in what we hope will be a new 'biological age' replacing an older, obsolete mechanical model that accompanied twentieth-century architecture.

The evolution of larger buildings

Although from the earliest days of civilisation there have been buildings for non-domestic use – temples, churches, castles, mills, warehouses, etc. – it was following the Industrial Revolution in the nineteenth century that new building types became widespread. At first some were little more than scaled-up houses. At the grander end of the scale, the Reform Club (1837) in London, for example, designed by Charles Barry, is an adaptation of an Italian Renaissance *pallazzo* with central courtyard (itself derived from the Roman *atrium*). Others were transformations of more indigenous types, such as the monastic cloister adapted for the Oxford Museum (1854–55).[1] Both these buildings glazed over the former open courtyard to transform it into a useable interior space, in effect making it possible to go beyond the two- or three-room deep principle. The modification to an inherited built form was made possible by new or improved technologies. The development of iron structures and the possibility of a fully glazed roof demonstrated in late Georgian and early Victorian glasshouses, for example, transformed the potential use of space.

Despite having no ventilation for the central, glass-roofed area, the Oxford Museum maintains a reasonably comfortable temperature even on a hot day (Fig. 4.1). The thick stone and brick walls that surround the court seem to have sufficient thermal mass to cope with summer heat. Tall, steep-pitched roofs on the buildings that wrap around the court help by shading the glass roof for much of the day. The large volume in the lofty, top-lit museum area allows hot air to rise above the zone of human occupation and loosely fitting glass roof tiles allow some air to seep out. The section was the principal design tool used to transform the traditional conception of rooms into the new, top-lit space of the museum.

Similarly the section best explains the chemistry labs (separated from the main building because of health fears associated with experiments) for the Oxford Museum, which was the University's first building dedicated to science. The architects Deane and Woodward adopted the form of the Abbot's kitchen at Medieval Glastonbury where the space is shaped as a funnel which is expressed in its steeply pitched octagonal roof capped by an open-vented lantern (Fig. 4.2). Originally designed to vent heat, steam and smells from cooking, here it allows noxious fumes to rise and be dispersed. Tall chimneys at each corner provided further ventilation located directly over workbenches. A typically nineteenth-century Gothic Revival building, the chemistry labs are also an expression of its primary function, although its appearance is more clone than adaptation.

4.1 The Oxford Museum, by Deane and Woodward.

4.2 Chemistry labs at the Oxford Museum.

However, lighting rather than ventilation was the major concern at the Oxford Museum. Lighting is largely related to surface area and the amount of light that can penetrate an interior is partly governed by the law (referred to previously) whereby the ratio of surface area to volume decreases with increasing size. This is discussed by Stephen Jay Gould, where he shows how a large cathedral such as Norwich can not simply be an expansion of a small medieval parish church, because the relatively smaller window area would be insufficient to light the increased interior volume (Fig. 4.3). The effect of a deeper plan on lighting the interior comes in his discussion on size and shape of animals. His analogy is based on the biological fact that some simple creatures never evolved internal organs because their food and oxygen can penetrate directly from the surface to all parts of the body – as we saw in Chapter 2. Gould makes an analogy with tapeworms, which can grow to 20 ft long but never exceed a fraction of an inch thick. 'Medieval churches', he explains, 'like tapeworms, lack internal systems and must alter their shape to produce more external surface as they are made larger.'[2] Norwich Cathedral is very long – 450 ft – but only 70 ft wide.

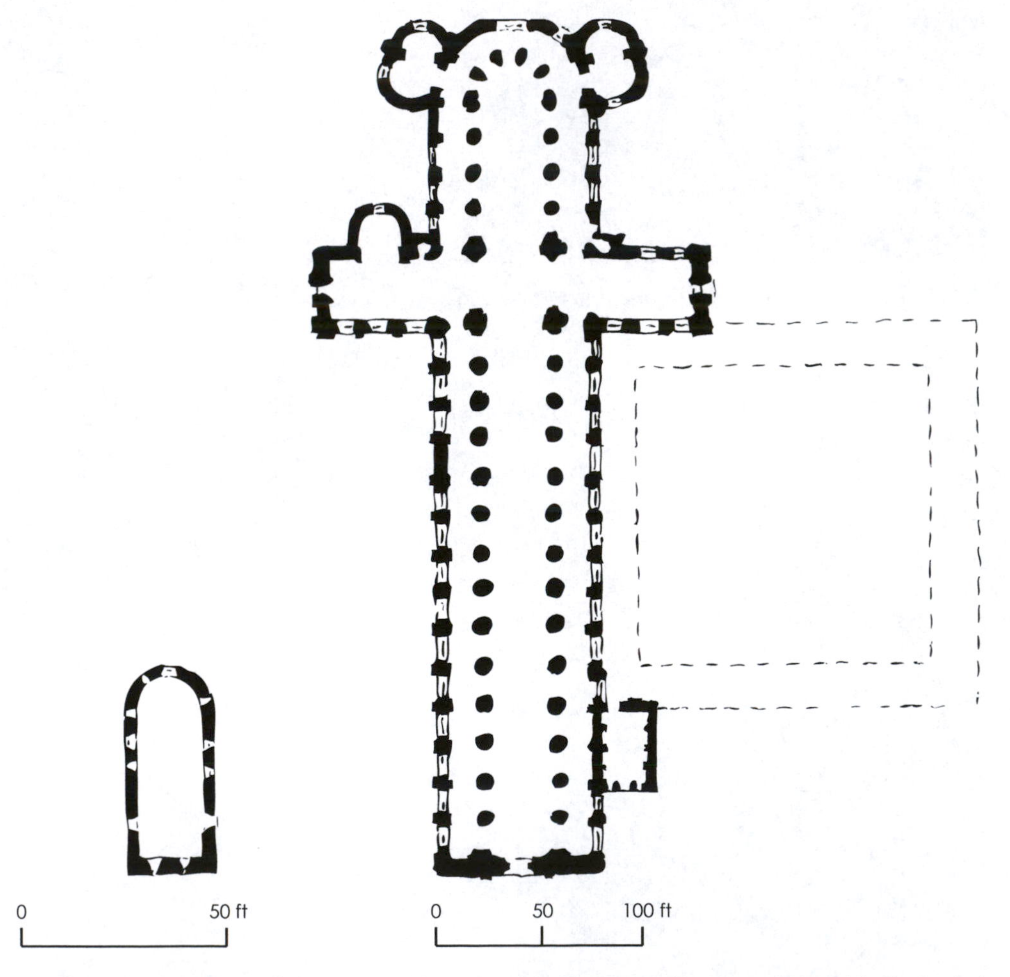

4.3 Comparative plans of the small parish church at Braxsted, Essex, and Norwich Cathedral. (Note that the small church is drawn at double the scale of the cathedral.)

Medieval cathedrals evolved their long thin shape because they were built, he continues, 'before the invention of steel girders, internal lighting and air-conditioning (which) permitted modern architects to challenge the laws of size ... No large Gothic church is wider than long; no large animal has a sagging middle like a dachshund.'[13]

He illustrates his example with plans, but his argument about size and shape, volume of space and wall surface, is given greater point by looking at a section (Fig. 4.4). For an obvious governing factor in the width of any medieval building was how far a timber truss or stone arch could span, which is rather small compared with modern steel or concrete structures, as he suggests. In addition, as the span increases, the weight and outward thrust from an arch upon the wall below become greater and the wall would topple unless contained by sufficient mass (as was the case in the Roman Pantheon, Fig. 7.10) or supported by buttresses. Transmitting these forces to the ground meant that window openings had to be relatively small. Medieval masons responded to these factors by the ingenious invention of flying buttresses that transmit the outward thrust onto adjacent and lower walls, piers or columns. Transferring some of the mass supporting the weight from the wall and roof above to flying buttresses has the additional benefit of allowing bigger openings in the wall itself. A typical Gothic cathedral has an aisle either side of the central nave. The judicious development of the section and high-level clerestory windows allowed light to be introduced into the centre of what is a relatively deep space – notwithstanding Gould's remarks – for the nave in some cathedrals is separated from the perimeter wall by two side aisles.

With the Gothic cathedral we see the development of a building type shaped by the use of section in relation to the properties of stone and ideas about light (see Chapter 7 for more on this). But integrating this with the plan and a consideration of the form creates its overall coherence. The nave section is contained at the west end by the twin towers that flank the entrance portal (itself an adaptation of the Roman triumphal

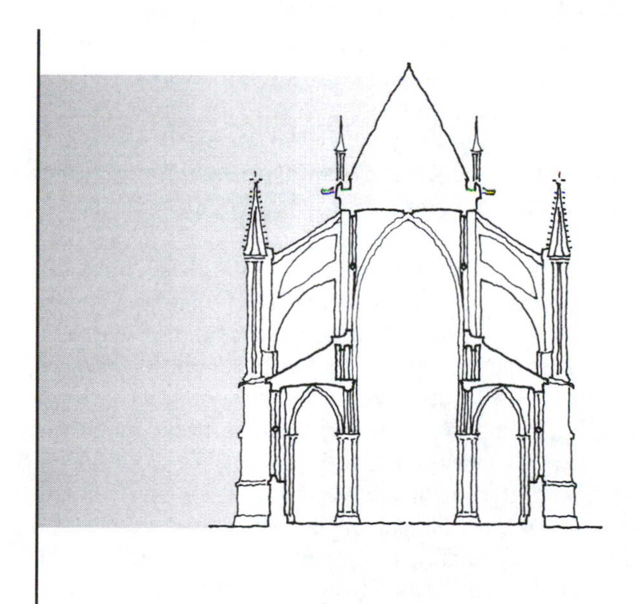

4.4 Typical cross-section of a Gothic cathedral.

arch. The Early Christian church adapted the form of the Roman Basilica and the Gothic evolved from Romanesque by transforming the nave with its hole-in-the-wall windows into a screen wall of coloured glass). The side aisles typically continue and join up as an ambulatory that wraps around the apse-ended chancel at the east end of the cathedral.

The Oxford Museum, a new type of building, employs the Gothic cathedral section of nave with two side aisles – although with an iron structure – and achieves an overall coherence largely because this new desire for a top-lit large space is contained within the surrounding walls of the cloister. These examples serve to remind us that successful architectural design depends upon achieving coherence through the combined consideration of plan, section, space and overall considerations of form. An environmental architecture goes further and develops this in conjunction with the advent of the building's surface as a potential source of energy.

The subtle development of the Gothic cathedral section indicates the significance of light in determining the morphology of building (and cities, as we will see in Chapter 8). The earliest buildings drew their light from a window in the wall or a smoke hole in the roof. The development of rooflights introduced a new possibility in built form. With the invention of iron structures and fully glazed roofs – pioneered by the Reform Club and the Oxford Museum – a third basic form was added to the morphology of buildings whereby daylight could be transmitted to the centre (Fig. 4.5). This will be discussed further in Chapter 7.

The importance of getting sufficient natural light became somewhat sidelined with Edison's invention of the incandescent electric lamp in 1879. This made it easier for environments to be artificially lit, and with higher levels of illumination than were possible with gaslight, oil lamp or candle. The first central power station and distribution system of electricity came into operation in 1881 in New York. From this time electricity began to be introduced rapidly into the office, factory and home.

In his book *Mechanisation takes Command*, the architectural historian Sigfried Giedion identifies the significance of electricity, and in particular the electric motor, in the development of the modern environment. 'It meant to the mechanisation of the household what the invention of the wheel meant for moving loads',[4] he wrote. A slow incubation period followed its invention by Faraday in 1831 but by the 1890s small and useful electric motors became available and entered the household, soon giving us the refrigerator and the vacuum cleaner, followed by washing machines, tumble dryers and dishwashers.

Electricity generation and electric motors played a crucial role in the development of the skyscraper in the 1890s. Electric lifts, heating pumps and power for lighting in conjunction with the development of the steel frame enabled buildings to become ever taller. Mechanical ventilation and electric lights made it possible for buildings to become bigger and less dependent on their envelope for their internal environment. The rapid development of the steel frame in Chicago transformed not only the appearance but also the thermal performance of buildings. Openings in load-bearing walls were essential for light and ventilation. These had tended to be small, as we have seen, because of a limit to the weight that could be carried by the wall material itself. The steel frame was quickly followed by the development of the glass curtain wall, which further reduced the weight but also diminished the building's thermal mass. This had the effect of removing the wall itself as an aspect of environmental design, in the traditional manner where its thickness and a balance of wall and window helped regulate the internal temperature.

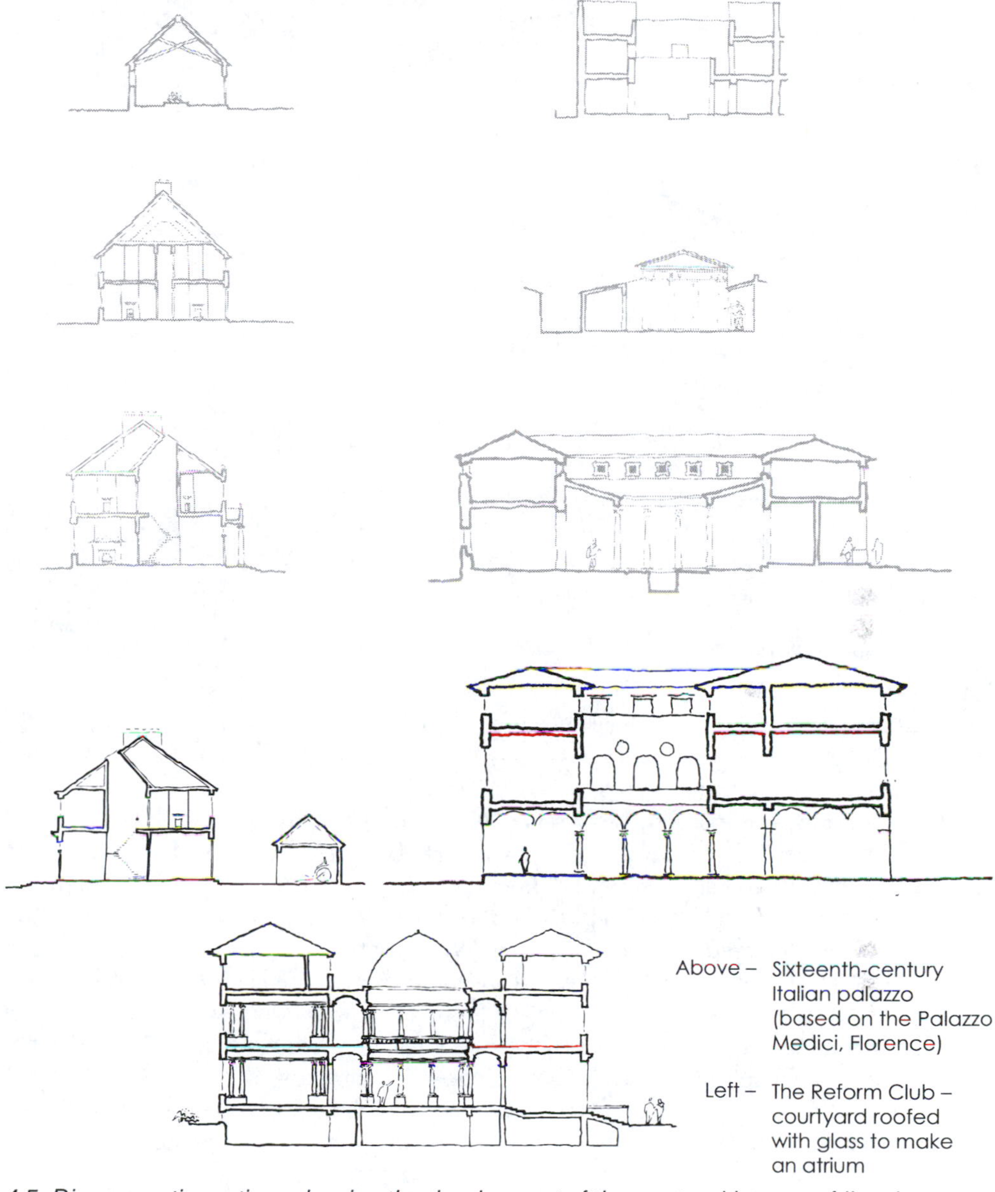

4.5 Diagrammatic sections showing the development of the courtyard into a roof-lit atrium.

Behind the tall and broad façades of early skyscrapers were light wells that also provided ventilation (Fig. 4.6). Essentially a contraction of the courtyard, these became smaller and smaller until no more than a ventilation shaft in effect, an arrangement that lent itself to becoming the vertical core of a mechanical ventilation system. Deeper plans and their accompanying labyrinths of horizontal and vertical ducts for services were developed. Heating, cooling, ventilation and lighting could now all be done by mechanical and electrical means. The advantage of filtering out the noise and dirt from the metropolis, combined with economic pressures of central city land prices and the cultural ambition to build tall buildings, led inexorably to the artificially lit and

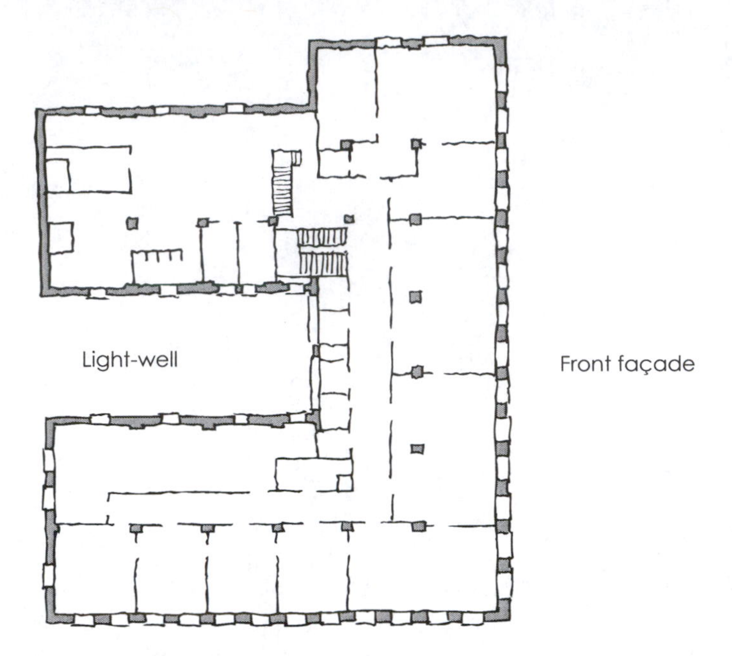

4.6 Typical floor plan of the Wainwright Building by Louis Sullivan.

air-conditioned buildings so familiar today. Following the model of Mies van der Rohe's Seagram Building, these may have one, two or even three complete floors given over to air-conditioning[5] (Fig. 4.7).

The importance of morphology to sustainability is seen clearly in the deep-plan and often fully glazed office buildings found in all major cities. In spite of the abundant perimeter glazing, artificial light is needed over most of the floor area. The extensive use of electricity for artificial lighting makes for a major contribution to CO_2 emissions (despite cooler fluorescent bulbs made widely available from the late 1930s).[6] It is estimated that 30 per cent of total electricity consumed in commercial buildings goes to artificial lighting.[7] In addition, the heat given off is normally dealt with by mechanical ventilation and cooling, which make further significant contributions to CO_2 emissions. These are direct results of the evolution of deep-plan buildings.

Much recent innovation in sustainable architecture has been to re-invent the wall as an agent of the processes involved in environmental control. The optimum proportion of glazing can be calculated to strike the balance among advantages such as solar gain and disadvantages such as heat loss and glare. Screens and blinds can be added for fine-tuning. The possibility of making an active 'skin' using advances in technology is an important consideration that will be discussed in later chapters.

The machine aesthetic and the sealed environment

The modern, glass, curtain-walled building type found its intellectual rationale in Le Corbusier's 'Five points for a new architecture'.[8] His 'domino house' – concrete columns pulled back from the façade supporting a flat concrete floor slab – allowed him to postulate the virtues of the 'free plan' and the full-width strip window or free façade over its hole-in-the-wall predecessor (Fig. 4.8). On the one hand this advanced

4.7 Office building in the City of London inspired by Mies van der Rohe's Seagram Building.

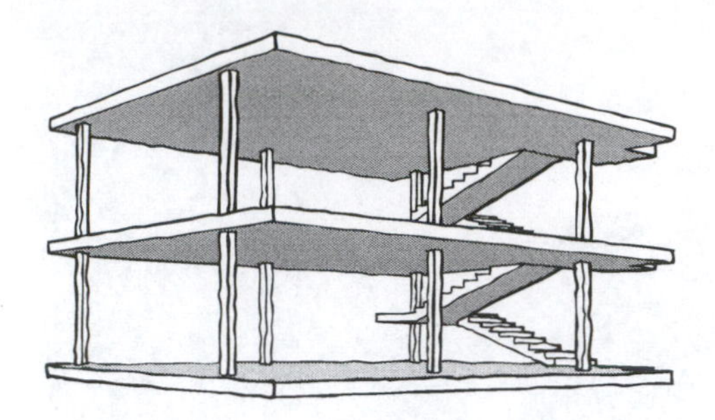

4.8 'Domino house' (after Le Corbusier).

the open, spatially flowing character of modern buildings, but on the other, the free plan encouraged the development of a naive functionalist approach that emphasised efficient internal relations, in particular circulation. (We will see a similarly impoverished functionalist approach in the development of the twentieth-century city in Chapter 8.) In conjunction with air-conditioning, this led to a situation where the traditional concerns about the environmental relationship between inside and outside could be virtually ignored.

Le Corbusier's houses, as we have seen, had windows and the additional benefits of expansive balconies and roof terraces redolent of ocean liners. But the hostel for the Salvation Army (The Cité de Refuge, 1929–33, which he designed with his broader ambitions for the city in mind) illustrates how he broke with traditional means of environmental control. Le Corbusier advocated what he termed 'man-made weather'.[9] His call for a machine age architecture was in response to the possibilities presented by technical developments. It coincided with the pioneer aviator Antoine de Saint-Exupéry who described how flying the fragile and open early aeroplanes engaged him more with the elements. That was a relatively innocent age of the machine before there was an environmental crisis, in contrast to the present time when enormous fossil fuel consumption is producing the CO_2 and other greenhouse gases that cause global warming. In *The City of Tomorrow* (1929), Le Corbusier lambasted the 'smelly streets, alleys and courtyards' of the historic city, and the arrival of the car added to such a view. We will return to this in Chapter 8.

The smooth flank of its original glass wall and pointed prow-like form gave the Salvation Army Hostel the appearance of a ship. Although this made for a novel form of building its meaning could be grasped by this visual analogy. It has a rather awkward relationship to the street, but Le Corbusier certainly intended this for it helped the building look more like a vessel, self-contained and independent from the fabric of the city.

Le Corbusier's original design incorporated a double glass 'neutralising wall' – sometimes designated a 'breathing wall' – a double skin between which hot or cold air could be blown.[10] But for reasons of cost, a single-sheet glass wall was built instead. Although mechanical ventilation was incorporated, this pioneering sealed building was not a great success – complaints were made about internal temperatures that reached 30–33°C and the lack of air. The vast, south-facing expanse of fixed glazing made this inevitable. With no opening windows the occupants were dependent on what proved to be an inadequate ventilation system for the high solar gains experienced. Windows had to be inserted, to Le Corbusier's disgust, and eventually

4.9 The Cité de Refuge (Salvation Army Hostel), Paris, by Le Corbusier.

the glass wall was replaced with a wall that embraced his later, more successful development of the *brises-soleil* (sun breakers, a form of sun control), of which more later[11] (Fig. 4.9).

The entrance sequence to the Cité de Refuge is designed to evoke the experience of boarding a ship by gangplank. The 'down and outs' of George Orwell's Paris and others in need could find shelter in this 'ship of salvation' – a deeply Christian symbol as well as one evoking Le Corbusier's technical and social vision. After 'embarkation' was complete in the front rooms of the building, the homeless climbed a stair to enter the dormitory block. On the landings – one stair for males and one for females – stand two great circular columns like sentinels. These are, in fact, the boiler flues coming up from the basement. Forced by its position in the middle of the stair, the homeless would pass close to them and, Le Corbusier hoped, would feel a welcoming warmth as they entered the sheltering body of the building.

A sailing ship designed and made to respond to water and wind is one of the loveliest of human creations. Sailing ships have to strike a delicate balance between resisting one element – water – and being open to another – air. The elegant form of a ship is derived in part from its design for reducing the resistance to moving through water. Steam ships, however, needed primarily to respond to water and could be largely closed forms. The development of ocean-going liners brought about open decks for passengers to enjoy the pleasures and health benefits of sea and sun, a development that we saw Le Corbusier exploit in his early houses. Aalto's early essay in International Style modernism, the Tuberculosis Sanitorium at Paimio (1928–33), can also be likened to an ocean liner (Fig. 4.10).

But the Modern Movement's fascination with the modern engineered forms of ship, aeroplane and car played a part in leading to the sealed building which mechanical services had made possible. Rejection of historical built forms by the early twentieth-century avant-garde produced a situation diametrically opposed to transitional buildings such as the Oxford Museum. There, processes of adaptation of existing forms to accommodate new functions and technologies produced new, larger forms that were

4.10 Tuberculosis Sanitorium at Paimio, Finland, by Alvar Aalto.

recognisable and whose meaning subtly changed, yet remained, within the realm of building but not machines.

Le Corbusier's pioneering experiment in designing a building as a machine-like vessel has left a large legacy. A common theme of some contemporary 'signature' architecture, it results in the appearance of buildings as sealed vessels that can appear disturbingly alien.[12] Drawing their imagery from space travel and science fiction, vessel-like buildings appear marooned in the landscape and imply an inhospitable environment in which no relationship is possible between inside and outside. This could be considered the antithesis of one of the arguments advanced here; that buildings should be more like biological organisms that interact more with their environment than machines.[13] The polluted air of cities makes this more difficult, of course, than in the natural surroundings where buildings first evolved.

Recent studies have shown extensive occupant dissatisfaction with sealed, air-conditioned and artificially lit offices. For a long time the Seagram Building (1954–58) in New York represented the epitome of elegance in modern architecture. Its spare, refined detailing and well-proportioned arrangement of columns, mullions and glass curtain wall was emulated all over the world. Many still find great pleasure in the grid-like streets of Manhattan, but critics have pointed out how the glass-walled skyscraper became adopted as the headquarters of corporations and see this kind of architecture as representing resource and energy-hungry global capitalism.

For some, the grid-like, curtain-walled skyscraper symbolises the twentieth-century world-view, an era when mechanisation took complete command. The geographer Richard Sennett (himself a lover of New York, as he describes in his book *The Conscience of the Eye*) argues that the gridded plans of American cities, like the grid-lines of a map, were essentially instruments for taking control of the natural world.[14] Particular qualities of a place were over-ridden by the desire to exploit natural resources. In his essay *The Question Concerning Technology*, Heidegger introduces the concept of 'enframing' to describe the insatiable demands industrial production puts upon the natural world.[15]

The elimination of recognisably human-scale elements, as in the glass-walled building, is a factor architects and engineers should keep in mind as they wrestle with the technical aspects of environmental design. Stripping away elements such as doors and windows – which express interaction and human control over their environment – makes it difficult for people to measure themselves against the built environment. Building façades are an essential part of the building's aesthetic and operation. As well as being the visible face of the building to the outside, the façade is the boundary between the internal occupied and conditioned spaces, and the external environment. A face has its eyes, mouth and nose to interact with the environment.

The early skyscrapers of Louis Sullivan, in contrast, have a distinct form: base, shaft (tying together many floors of windows) and top. This form was derived from a classical column, which in turn was based upon the tripartite nature of the human body with its feet, torso and head (Fig. 4.11). Sullivan also gave detailed attention to the façade at street level, surrounding shop windows with ornament and introducing a mezzanine to break the eye's sweep upwards. This approach, which became a typical feature of early-twentieth-century American tall buildings, recognises the need to provide visual interest in the appropriate place at the appropriate density to satisfy the human eye.

In the more innocent *Architecture of the First Machine Age* – to borrow the title of Reyner Banham's book – mechanical services were considered to be more reliable and particularly beneficial in cities due in part to their ability to control pollution and noise.[16] If external noise and pollution are high it is not easy to ventilate by natural means, of course. But by using devices such as acoustic attenuators and with careful positioning of air inlets and controlled movements of air through a building, this can be achieved – as we will see in the next two chapters.

A major theme of Banham's later book, *The Architecture of the Well-Tempered Environment*, is the expression of services. He called his book an 'unprecedented history' of 'the modified environment' and he laments the curious omission of this important aspect from the history of modern architecture.[17] Louis Kahn's Richards Memorial Laboratories mark the first point, argues Banham, when the increasingly significant role of mechanical services is given clear expression in the form of a building. (Although he points out that the form and spatial arrangement of Frank Lloyd Wright's earlier Larkin Building is largely determined by its mechanical system of ventilation; Figure 6.10.)[18] He does criticise Kahn, however, for effectively putting the

4.11 The Wainwright Building, Buffalo, by Louis Sullivan.

services in 'cupboards'. He would have preferred a more explicit expression in glass boxes perhaps, or as tubes creeping across the surface of the building, as in the project he illustrates by Mike Webb of the then fledgling Archigram.[19] Expressing the mechanical services on the exterior of the building became a major theme of buildings in the age of cheap fossil fuel and Richard Rogers and Renzo Piano's Pompidou Centre (1971–77) is perhaps the most well known example (Fig. 4.12).

In a curious way, this approach to architecture is not unlike Robert Venturi's 'decorated shed' theory introduced in his book *Learning from Las Vegas* written with Denise Scott-Brown and Steven Izenour. But where his buildings were decorated with historical motifs and Pop Art-like wallpaper, the Centre Pompidou and others in this vogue use brightly coloured services and structure as a form of expression. This reduces the role of the building enclosure to a box that is mechanically serviced.[20] In contrast, so much of a traditional building's expressive character derives from how its form, mass, materials, organisation and elements such as windows – and in some cases doors – contribute to making a comfortable micro-environment at the same time as being related to human scale.

Mechanisation loses command

The environmental crisis that emerged in the latter part of the twentieth century points to mechanisation losing command. It has become evident that new approaches to architecture and urban design are required. The environmental impact of a building and in particular the CO_2 that it produces depend on its overall design. Natural ventilation and

4.12 The Pompidou Centre, by Richard Rogers and Renzo Piano.

daylighting, which reduce energy consumption, have an important impact on building form and appearance. These in turn are related to the need to control solar gains and the thermal mass of the building. A holistic approach is best, one where the built form and the engineering systems work together. Building design must also be consciously guided by principles of composition which might embrace elegance, rhythm, proportions, symmetry or asymmetry, balance, pattern, contrast, discordance, etc.

An early sign of a reconsideration can be detected in the very different way Le Corbusier designed the Unité d'Habitation (1946–52) ten years after the sealed vessel that was the Salvation Army hostel. Although more generally known for its social aspirations, the Unité at Marseilles also shows how natural ventilation, solar shading and thermal mass can work together in an environmental approach to dwelling. And where his early architecture aimed for machine-like imagery and the play of pure forms in light, this later work explored the interplay of light and shade. Le Corbusier considered this to be one of the essential mysteries of life – seen most strikingly in the phenomenon of germinating seeds where emerging shoots rise to the light and roots go down into the darkness.[21] The *brises-soleil* played its part in reducing solar gain but also casts shadows, emphasising the contrast of light and dark that became a central part of Le Corbusier's later symbolism.

The section of the Unité organises the accommodation so that every dwelling can have cross ventilation (Fig. 4.13). Each apartment has a double-height living room that opens on to a balcony shaded by a *brises-soleil*. The height gives more sky hence better daylight for the room and the *brises-soleil* prevents excessive solar gain. The slim plan of each dwelling running the full width of the building ensures that thick party walls are on both sides of most rooms. In conjunction with the concrete floor slabs, this gives a high thermal mass. The depth of the section ensures that the sun never directly affects some areas. The double height also helps to make the living rooms comfortable in this Mediterranean climate, by allowing hot air to rise above the zone of human occupation. The apartments have a balcony on either side of the building shaded by the *brises-soleil*. Here we see how his use of plan, section and materials

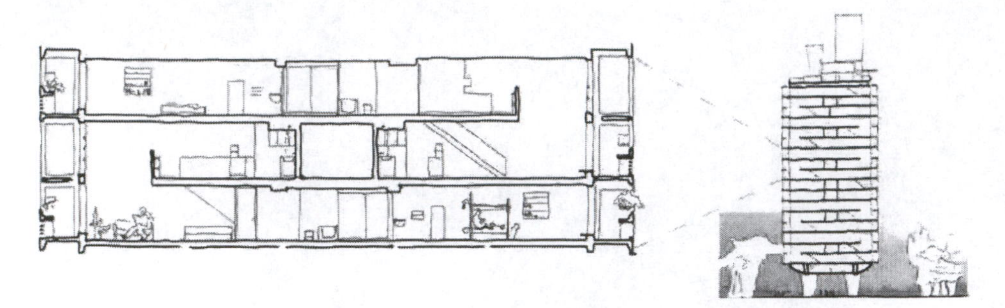

4.13 Cross-sections of the Unité d'Habitation, Marseilles, by Le Corbusier.

work together to create a comfortable environment taking account of thermal mass, solar shading and cross-ventilation. Banham wrote: 'the advantages of the traditional massive wall were argued back one at a time.'[22] Such considerations are important to a more sustainable architecture that depends less on mechanisation. Although most of the examples in this chapter are not really large buildings by contemporary standards, very few buildings before or since are bigger than the Unité.

In considering form and appearance, the idea of shelter, or its expression, will be different in large modern buildings from how it is in vernacular dwellings. With its façade of heavy concrete solar shading, Le Corbusier's Unité has been interpreted as making a 'battered bunker', a psychological protection for the post Second World War psyche.[23] Doors and windows – the elements most obviously human in scale – are drawn back within the section so that the building appears as a rugged, homogeneous mass with deep, unmediated openings. This was appropriately robust, he believed, to suggest cave-like shelter and also to operate at the scale of building-as-city (there is a shopping street in the middle and children can play on the roof terrace where there is also a running track). The inhabitants can withdraw to a low-ceilinged interior or sit out on a lofty, shaded balcony. In 'larger forms' such as this, the idea of shelter, or appropriate psychological protection, must deal imaginatively with the elements of building that humans identify with from their everyday experience.

Some recent sustainable architecture has incorporated similar devices for shading the wall in order to reduce excessive solar gain and eliminate the need for air-conditioning. The building's form and skin partially take the place of mechanical systems. New materials and operating systems allow sunscreens to be fixed on the outside of buildings, unlike their forebears in domestic buildings (although external shutters with moveable blades are commonplace on the Continent). Making use of advances in technology and computer operating systems, these can prove very effective in reducing solar gain, but do not always address the issue of human scale and user-friendliness in a satisfactory way. A good deal of post-war modern architecture (although, of course, not the best) was designed largely through the use of section. However, simply extruding a singular technical solution to the problem of solar gain, for example, can result in the building looking more like a piece of apparatus than a human habitation. In contrast, a sunscreen contained within a gable, for example, reads as a detail within a recognisable form (Fig. 4.14). Sunscreens with or without moveable blades, opaque and translucent photovoltaic panels for use in walls or rooflights are among many devices that have entered the market, all of which benefit from being properly integrated with the more traditional concerns of design. We need

4.14 Restaurant for the Aldeburgh Festival Hall, Snape Maltings, by Penoyre and Prasad.

reference points in the body of a building to understand its relationship to us in its fundamental role of providing shelter, not only for the physical requirements of bodily comfort, but also psychological needs.

With heat gain becoming a critical concern of larger buildings, some of the most inventive sustainable buildings of recent years have been those where enhanced or assisted natural ventilation has been pursued. Wind towers from extremely hot climates, such as the Gulf region can be an inspiration for contemporary sustainable design.[24] A more scientific understanding of airflow has been applied to vernacular buildings that evolved to make living conditions comfortable and helped develop principles for cooling modern buildings in the West. The addition of devices such as these to induce through draughts can add distinctive character. The loss of chimneys from modern buildings removed an element that added to the skyline. Not only did the chimney-piece provide a focal point to a room but also the chimney itself added something to the skyline. The impetus for incorporating wind towers in modern buildings came partly from the need to reduce energy consumption and the burning of fossil fuel, so it is pleasant to think that natural ventilation devices have reclaimed the skyline in the name of sustainability. The expressive chimneys of an earlier age symbolised the idea of hearth and home (as well as being symbols of status or power). These latter-day equivalents symbolise the role of built form itself becoming once again the modifier of the internal environment. Natural ventilation devices and wind turbines will help create the skyline of the future.

Revolving cowls turned by the wind occupy the top of some modern naturally ventilated buildings to enhance air extraction, such as can be seen in Glenn Murcutt's buildings, or on Bill Dunster's and Bio-Regional's BedZed. These remind us that some vernacular buildings of Britain developed devices to assist with ventilation. The oast houses of Kent are perhaps the most familiar and the most picturesque (Fig. 4.15). Their cowls that rotate in response to the direction of the wind create a pronounced wind effect which, in conjunction with the stack effect, draws hot air from a fire below, up and over the hops, drying them for use in beer production.

The name oast *house* itself is a reminder of how their appearance in the landscape is but a particular adjustment of domestic, and therefore, human scale. The white cowl is a distinctive element that caps a brick-and-clay-tiled building rising from the earth like the farmhouses of the region. In a similar way, traditional windmills draw much of their emotive appeal from the juxtaposition of a tower-like brick house with domestic scale doors and windows beneath ladder-like revolving sails. Like sailing ships, they are partly protective containers and partly open to the wind. Both these building types strike the eye, not only because of their unusual appearance but also because their wooden superstructures rise high over their surroundings to catch as much wind as possible.

4.15 Oast houses (above) and whisky distillery (below).

The same issues of scale, materials and appearance in conjunction with non-domestic function are satisfactorily and pleasantly solved in industrial vernacular buildings such as maltings and Scottish whisky distilleries. Pronounced roof vents cap the skyline yet seem to grow organically out of the roof and building below. They also speak directly of their detailed function within the whole as does a camel's hump or a giraffe's long neck.

At Snape, Arup Associates retained the basic form of the maltings' ventilators but enlarged them to cope with the demands of ventilating a large auditorium for the Aldeburgh Festival. The demand for efficient circulation in modern building types has contributed to the development of deep plans which has prompted the use of enhanced ventilation and where cowls have been effective in providing this. At the Benenden School, van Heyningen and Haward located tall chimneys capped with metal cowls at the back of double-banked classrooms to provide a stack effect and cross-ventilation (Fig. 4.16). This lends the design an echo of these earlier rural, industrial vernacular

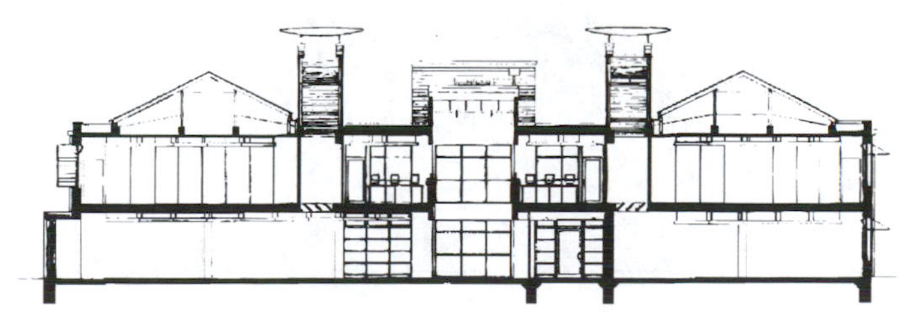

4.16 Benenden School, by van Heyningen and Haward.

buildings. Farms and barns evolved over hundreds of years but associated buildings such as maltings became scaled up in the nineteenth century to meet increased demand for beer. They were an adaptation of a long tradition, but the buildings were much larger using steel roof trusses and a new rationality informed the long rows of repeated windows. Yet the brick walls and pitched roof retained the memory of a deep past where vernacular buildings provided shelter, their dominating long roof-line was made more abstract under the thin sheen of slate tiles, and their strong profile was punctuated by ventilation cowls. The combination of size, abstraction, romantic profile and traditional protective forms and materials gives these rural industrial vernacular buildings an archetypal character that makes them good models for some locations. Jung's conception of archetypes carrying a deep, collective memory might be a good aim for architecture more generally. Feilden Clegg Bradley developed an interesting variation of ventilation cowls for the Heelis National Trust offices at Swindon (Fig. 4.17, and see Chapter 6).

The polluted air makes it more difficult to naturally ventilate buildings in our inner cities, as we have mentioned. In addition the extremely large size of so many modern commercial and institutional buildings increases the difficulty. In these circumstances – where air-conditioning is common – some mechanical assistance combined with the enhanced stack effect from chimney-like forms can be used for ventilation. A well-known good example of this is Portcullis House, a very large eight-storey building by Michael Hopkins and Partners (Fig. 4.18). Providing offices for Members of Parliament and their assistants, Portcullis House is located adjacent to Westminster Bridge, an area heavily polluted by traffic. Fresh air is drawn in as high as possible above the zone of air polluted by vehicular emissions through air intakes that wrap around the base of chimneys that cap the pitched roof. Fresh air is fed to underfloor plenums and then into the room where it rises as it warms and exits into ducts on the outer walls. These ventilation pathways run up the outside walls of the building and across the sloping roof to exit through large bronze chimneys that add interest to the skyline. Looking like a cross between steam train funnels and familiar chimney pots, these details, the scale of the building and its massing show a considered relationship to its neighbour, Norman Shaw's New Scotland Yard.

Short Associates' Coventry University Library is capped with what looks like scaled-up versions of Gulf *badgirs* (although the stack effect is used here rather than the wind's force), an expressive theme continued with variation in their more recent

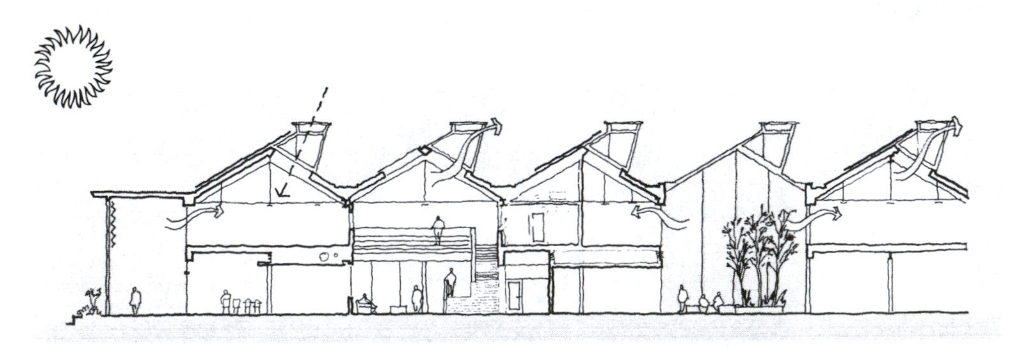

4.17 The Heelis National Trust offices, Swindon, by Feilden Clegg Bradley.

4.18 Portcullis House, London, by Michael Hopkins and Partners.

School of Slavonic and East European Studies (SSEES) for UCL. Where Kahn's Richard's Memorial Laboratories service towers disperse gases into the atmosphere, the similar-looking elements in Short's buildings are part of a strategy to reduce them (Fig. 4.19). The SSEES is a very large – 3,500 m^2 – seven-storey building that further illustrates the kind of problems faced by sustainable architecture in tight, dense inner city sites. Research indicates that London is warming rapidly and that CO_2 emissions will rise by 4.7 per cent by 2050 as air-conditioning increases.[25] The design team applied cutting-edge university research in applied mathematics, theoretical physics

4.19 The School of Slavonic and East European Studies, UCL, London, by Alan Short and Associates.

and experiments in fluid flow to implement passive down-draught cooling (PDC). PDC enables naturally ventilated buildings to be cooled and use less than half the energy of a typical mechanically ventilated building. The SSEES employs a ring of cooling batteries around the head of a central lightwell and the cooled air enters here on hot summer days to be driven by negative buoyancy through the building. Stale air is exhausted through a ring of vertical ducts some of which peep above the front parapet and look like *badgirs.* The street façade of this building – load-bearing brickwork expressing the stairwell behind – is sealed but its appearance acknowledges the rhythms and organisational principles of the Georgian street in which it stands. In this example we see how cutting-edge science and traditions of built form combine to great effect. Mechanisation is losing command as the building's form, fabric and details come to play a greater role once again.

Banham characterised the history of architecture as creating 'massive and perdurable structures'.[26] Very much a product of his age, the freedom seeking Sixties, he was critical of such structures, arguing that 'power-operated environments' were the order of the day and that architecture should be heavily serviced, lightweight structures and express this.[27] At the very end of his book, however, he made an unexpected retreat from this position. Curiously prescient of a theme pursued here – reducing energy use through the design of built form – Banham's final example is St George's School, Wallasey (1961), at that time a little-known example of using passive solar energy for heat – supplemented by electric lighting and the inhabitants. Not unlike Le Corbusier's original design for the Salvation Army building, it has a double-skin glass 'solar wall' but incorporates various mechanical devices for thermal control (Fig. 4.20). Thick brick walls, concrete floors and roof give the building a substantial thermal mass.[28]

Despite his focus on the expression of services in architecture, Banham concludes his discussion by pointing to how environmental science can be applied to traditional knowledge about a building's performance to make perhaps the most satisfactory environments for human beings. This, of course, echoes the suggestion made here of how to create comfortable environments while minimising the environmental impact. From the more biological perspective we have outlined, and taking into consideration the desire to produce well-rounded designs, criticism could be levelled at the form of St George's School being determined solely by the section and technical matters of

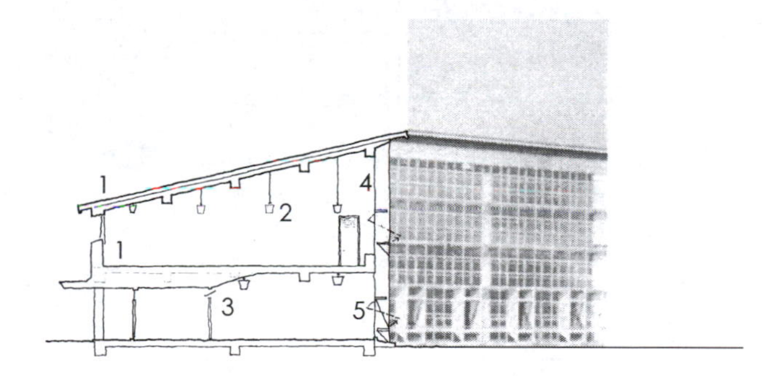

1. Concrete roof and floor giving thermal mass
2. Light fittings providing some heat
3. Ventilating windows
4. Double-skin solar wall
5. Adjustable ventilating windows

4.20 St George's School, Wallasey.

environmental control which produced a rather monolithic and unforgiving appearance.

Aspects of culture and tradition are to be valued and can form a background against which we can assess those aspects of technical progress that contribute to making an architecture for a more sustainable future. Although much twentieth-century architecture was driven by a belief in giving expression to the machine, there were other strands that can help us see ways to shape the future. In particular the works of Aalto were essays in how the aspirations of the Modern Movement could be combined with attributes of a particular landscape and cultural traditions, as we saw in the last chapter. Although not an overtly sustainable building (the environmental concerns of today were not apparent then), his town hall at Säynätsalo (1949–52) in Finland is an interesting example (Fig. 4.21). Concerned at the overwhelmingly large forms of building created for twentieth-century institutions, Aalto divided the accommodation into three elements that wrap around a courtyard. Turf stair treads rise to the central courtyard which suggests a melding of building and landscape. Raising the courtyard to the first floor effectively reduces the apparent height of building to a single storey, lower than a typical house (Fig. 4.22). Not only does this affect the scale of the building and its impact upon users but it also allows for more sun penetration into the courtyard. This enhances the micro-climate in this cold region. The building is a rugged, earthy brick on the outside but the courtyard walls are largely of glass with

4.21 Section through Säynätsalo Town Hall, Finland, by Alvar Aalto.

4.22 View of courtyard in Säynätsalo Town Hall.

thin vertical poles over window mullions that echo the surrounding birch trees. The council chamber has a roof structure that evokes the branches of a tree.

Aalto's is a good example of the approach advocated by Michael Dennis of combining the two traditions that we have inherited – the particular vernacular forms of a region and the Modern Movement.[29] As in his Villa Mairea, he rejected the machine expression of the early Modern Movement and turned to how Finnish farmsteads grouped buildings together, adapting and transforming these arrangements to form a more humane modern architecture rooted in tradition. He made extensive use of the indigenous timber that characterises Finnish vernacular, exploiting the technical advances in lamination and plywood that resulted from industrial production.

A consideration of the BRE Environmental Building will serve to summarise what has been outlined above and points the way to the more detailed discussion of environmental design in the following chapters. A late-twentieth-century example, it shows clearly and in general terms how an architecture based on inexpensive fossil fuels, sealed boxes, mechanisation and high-energy use was losing its role as a paradigm.

Figure 4.23a shows a section through the building. This is derived to a large extent from consideration of what might be called a golden triangle of sustainability: control of solar gain, thermal mass and ventilation. Controlling solar gain is discussed below. Thermal mass is provided particularly in the exposed concrete soffits of the ground

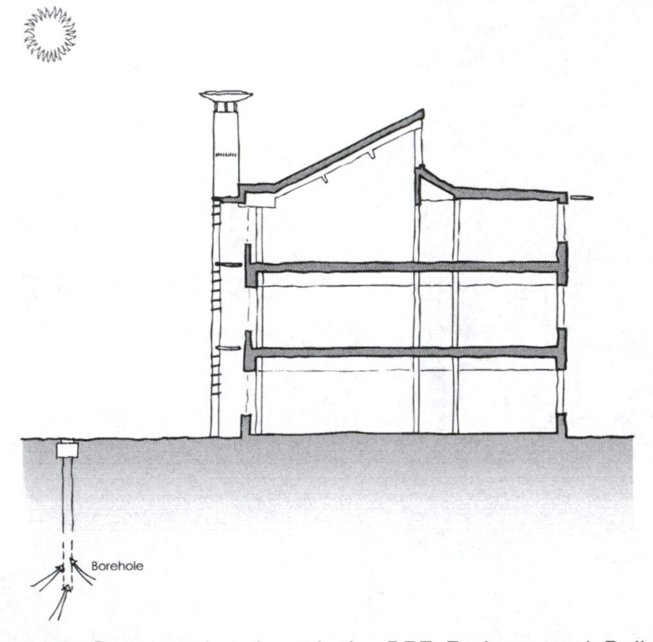

4.23a Cross-section through the BRE Environmental Building, Garston, by Feilden Clegg Bradley.

4.23b South façade of BRE Environmental Building.

and first floors and the heavy timber roof of the top floor. Efficient ventilation is achieved by ensuring adequately sized air paths and by incorporating night cooling (all discussed in subsequent chapters). Note that the use of cooling systems based on natural ventilation and thermal mass has a direct effect on the morphology of the building. It tends to produce plans with ample access to the perimeter, as is the case here, or spaces that can be ventilated using the stack effect with large air inlets and tall towers.

The section in Western architecture has an interesting history. It was during the Renaissance in 1515 that Raphael was appointed by Pope Leo X to produce a set of drawings for the buildings of ancient Rome. For what may have been the first time, each building was to be represented by a plan, elevation and section rather than perspective drawings. It may be surmised that the importance given to the section represented a new way of thinking about building – one that was more physiological and concerned with function. This is probably related to the rebirth of anatomy during the Renaissance and with it the recognition that dissection was required to understand function. Leonardo da Vinci, whose beautiful anatomical drawings married art and science, dissected more than 30 corpses. Both the words section and dissection come from the Latin *secare*, which means 'to cut'. The section is the privileged cut of the architect, opening up the building to understand how it works in a similar way to how a pathologist examines a body, although here it is to ensure that the 'body' will be alive and well in translating drawing to building. This way of thinking, understanding and then designing accords well with our views of a sustainable architecture that sees function in a context of history and aesthetics.

By dealing correctly with solar gain, thermal mass and ventilation in a number of temperate climates – including the UK – it should be possible to provide summertime comfort without the use of any significant amount of cooling (see below). An abundance of natural light controlled to avoid glare will reduce both summer and winter needs for artificial lighting. Proper levels of insulation will minimise the need for winter space heating – demand reduction is as always the starting point for sustainability.

Figure 4.23b shows the south elevation of the Environmental Building. The five ventilation stacks act as a new massive order that might be thought of as adding to the Doric, Ionic, etc., and which might be called the Environmental. They provide both variety by interrupting the horizontal lines of the solar louvres and add a calm stability by seeming to stake the building to the earth.

Approximately 50 per cent of the south façade in the office area (to the right of the door in the photograph) is glazed. This reflects a balance (and a sustainable architecture will have many) among heat loss, the need for daylighting and passive solar gain and the avoidance of glare and overheating.

The roof allows for the evolution of the building and it illustrates a significant difference between biology and designers. The design permitted change anticipating that the building would become more sustainable and accepting that it would take some time. This is a luxury (and in a world of rapidly growing awareness of the effects of global warming one that may not long be available) that biological organisms do not have. They need to be sufficiently adapted for the present or risk perishing,[30] an important concept for sustainability which is not given enough attention. Consequently the larger south-facing area is set at an almost optimum angle of about 30° for the performance of photovoltaic panels. (A smaller area of PV panels can be seen in a less optimum but more visible position on the left of the photo.)

The 'skin' of this building is an unusual mix of old and new – a combination of recycled Cambridgeshire bricks, state-of-the-art PV panels and adjustable translucent glass louvres, which control solar gain as required. This is a very responsive façade, capable of making subtle adjustments to varying light and temperature conditions. And it is a 'layered' façade with different levels of environmental control producing a visually rich architecture.

To deal with the high heat gains in the conference room (about 100 W/m^2), a borehole was drilled about 70 m down into the chalk below to withdraw water at 12°C to provide cooling; a second borehole is used for discharge. Thus the building is firmly located in a particular place with a functionality that depends on the environmental assets ranging from 'earth to sky'.

This also reminds us that much of engineering design and biology is based on taking advantage of differences, between the ground and the air temperature inside a building, for example, or between the core temperature of a mammal and the air temperature around it.

The keys for an environmental approach to architecture are to understand the important functional considerations in order to develop new forms (or at least variants of the old ones) and then to ensure that they please us aesthetically and are meaningful, giving us an architecture that enriches our lives. In a sense the question to keep in mind at all times is what is architecture for? This is a large question, of course, and will vary from individual to individual and culture to culture. Our own preliminary or outline answer might be that good architecture responds to Eliade's dictum, discussed earlier, that it is 'the universe man creates for himself'. At its best, architecture embodies our thinking about the world and helps us understand our place in the world. It reflects our relationship to the world through the way it is made. In addition it supports our inner selves by responding to our psychological needs.

Materials, new skins and future forms

We might begin this section by asking what we want materials to do and how we select them? This is a large and open question. For the great majority of humankind's time on Earth, there has been little opportunity to choose. Generation after generation lived in the same small worlds and learned to use materials that were ready to hand. Trees provided timber, stone could be quarried and cut into blocks, earth or clay could be mixed with water and baked in the sun or fired to make bricks, and reeds or straw could cover a roof. From this basic palette of materials came an enormous variety of vernacular buildings that reflected the different properties of wood and stone, climate and cultural traditions. In addition the fact that these buildings were made entirely by hand affected not only their scale but also the feel of the material because the marks of the craftsman were visible.

Buildings, like other material artefacts, became the product of a particular craft – or a combination of crafts – and more and more refined techniques evolved. Hundreds of years of accumulated experience led to economic building methods using the locally available materials to produce built forms that gave a comfortable internal environment. These traditions were inherited and localised. Alec Clifton-Taylor's book *The Pattern of English Building* mentioned earlier shows the extremely sensitive nature of vernacular responses to regional variation in climate and available materials in a single country. How refined these adaptations to a place became is conveyed in the title of

Bernard Rudofsky's book which he calls (with no qualifying adjective such as vernacular) *Architecture without Architects.* The enduring appeal of vernacular building lies in the organic relationship it shows between particular places of the Earth, its climate, its materials, topography, etc., and how mankind has adapted to it.

With ostensibly cheap transport (the environmental impact is rarely costed) and vernacular traditions no longer alive and well, our choice of materials is virtually unlimited. Therefore, over and above the general criteria of cost, stability and fitness for purpose, sustainable architecture needs to give particular consideration to the impact materials have on the broader environment. Also the effect they can have on human health and well-being must be taken into account.

Very broadly, there is a similarity between nature and engineering (and by extension, architecture) based on energy, and thus, materials. Plants and animals, in order to survive, tend to develop structures and physiologies which optimise the use of the available resources. We are really only at the beginning of the process of developing and selecting building materials and engineering systems that work as hard as natural ones. But under the stress associated with global warming we are starting to see more efficient forms, materials and systems that will be more likely to survive.

Material selection is a complex field. There are a number of factors involved in a material's impact upon the eco-systems of the planet: extraction of raw materials, pollution associated with manufacturing processes, transportation, and energy used in converting raw materials to building components. Issues of transportation are not new: flint implements were imported into present-day Cyprus as early as 12,000 years ago.[31] Materials selection has always had an element of the non-local. On the other hand, frugal Zen monks in reacting against a gardening tradition of using stones imported from great distances at high cost advocated the use of local rocks.

Consideration should be given to materials that have low embodied energy, construction that is low on waste, buildings that can be taken apart and recycled at the end of their life, strategies that are now familiar in architectural practice. It is important to get the materials right and to construct the building in an appropriate way. The more general interest in what is called 'local economies' might be kept in mind and pursued for building materials as well as the more specific architectural aim of making buildings 'carbon neutral'. Local materials can also help prevent an area feeling like it is the same as 'everywhere else'. The social implications of materials must also be considered, for it is unacceptable, of course, if local people's lives are destroyed to provide them.

A number of materials used in buildings are very energy-intensive (i.e. have a high embodied energy) including cement, steel, glass and aluminium. Aluminium production (for all uses), in fact, consumes 5 per cent of all Chinese power.[32] For comparison, in the UK aluminium needs approximately 30 times more primary energy than bricks and 100 times more than timber. What is important though, is to look at whole systems and whole building components. It is not just the energy input into the individual items but the assembly as a whole that needs to be compared. For a house wall, for example, one could compare masonry construction with, say, a steel frame and infill panels from the point of view of embodied energy. In the UK the energy for material transport could be significant hence the current importance being given to sourcing materials 'locally'. Brick gives good thermal mass but would involve a much larger quantity of material and labour than metal in construction. In some parts of the world it is possible to build simple one-storey earth-bag buildings with materials taken directly from the site.[33]

For buildings up to three or four storeys timber is often an excellent environmental choice (provided you ensure that it comes from a renewable source). There may not be enough timber to build a world out of it but it has enormous potential. Other biological materials are very promising. Plants use land more efficiently than animals and cellulose-based insulants will become more common. (However, there is also a case to be made for Scottish sheep, for example, scavenging on steep slopes covered with heather and providing wool to be incorporated into the palette of building materials.) Commercial photovoltaic panels based on organic materials are also predicted for the near future. Indeed, when one looks at the range of possibilities, this century may turn out to be a 'golden age' where biology and, perhaps, especially botany, are instrumental in everything, including the climate of cities, the construction of green walls and roofs, the internal environment, the treatment of water and the production of energy. Surely, more examples of mechanisation losing command.

Something the designer should keep in mind when wrestling with the quantifiable factors bound up with specifying materials is the impact they have directly upon people's feelings about buildings and spaces. We describe materials as feeling warm or cold. There is often a scientific basis to this in that metals, for example, are good conductors of heat and therefore take heat quickly from the body. Wood, in contrast, is a poor conductor and hence feels relatively warm to the touch. But compared with a woollen carpet, a wood floor feels cold to the feet. These physical sensations probably correspond to psychological states, such that a timber-panelled room would feel more warm and comfortable to inhabit than a predominantly metal one, just as we say that one colour is warmer than another. There is another, physical, reason for this, in that a polished metal surface will emit less radiation than a wooden one at the same temperature. Recent research in Japan has highlighted the psychological importance of materials – people who look at hedges experience a feeling of relaxation while concrete fences produce stress.[34]

Materials can also convey a deeper meaning. Our word material, for example, is derived from the Latin *materia*, which means the hard part of a tree. The word *materia* is closely related to the Latin *mater*, which means mother. The trunk of a tree produces branches or offshoots just as a mother produces offspring. So our word material has its root in reproduction, the eternally regenerative power of Mother Nature. It is interesting, if distressing, to reflect upon how far our modern materialistic world-view has departed from its meaningful root. But perhaps the enduring appeal of wood as a material for building lies in an intuition of this. We live on 'Time's Line' as we proceed from cradle to grave, yet we live within Nature's 'eternal return' – 'Time's Cycle', spring, summer, autumn, winter, spring – as most obviously seen in the 'rebirth' of trees each spring. The grain of wood can remind us of this. Knots and worm-burrowings may be accepted and acclaimed in Japanese timber-framed houses where the wood is left planed and unpainted.[35] Semi-translucent paper screens show threads and fibres that remain unpulped in the papermaking process. This approach shows the 'life' of the material, its growth and the processes involved in transforming it from one state to another. As we saw in Chapter 2, trees take in carbon dioxide from the atmosphere, so using wood from a sustainably managed source for building has a direct practical benefit. Timber used in building may not decay for hundreds of years – thereby storing CO_2 it has taken from the air – and recycling will help make the best of this exceptional resource.

The meaning contained in material can shift over time, but natural materials foster the sense of an organic relationship between human habitats and the broader environment. For example, stone became associated with eternity from the earliest days of mankind, standing forever outside of the cycles of growth and decay. Bone became linked to stone as being the part of humans that appeared to resist decay in a similar way. Hence, perhaps wood, with its marks of growth, its regeneration each spring, became used for coffins (and hope of regeneration) and the gravestone fixes its mark forever. Granite is often used for headstones because of its extreme hardness. Today polished granite is a popular cladding material for buildings although many architects prefer a stone that shows its age by weathering, or sandstone that shows its origin in sedimentation, or Portland stone in which fossil life is visible.

In the nineteenth century the new science of geology revealed the aeons of time over which the Earth had evolved. The discovery of fossils clearly dating from much, much earlier than the Biblical account of Creation, caused profound anxiety and doubt. Science and technology were seen to be at odds with faith. Thinkers such as Carlyle and Ruskin considered that industrial production, new technologies and the scale of industry amounted to an assault on Creation itself. The 'dark Satanic mills' blackened the countryside, railways were driven through cities, building materials could be moved anywhere. An organic order from time immemorial in which humans were symbiotically immersed – through their labour and the things they produced – was being destroyed. Stone, the most noble material of early architecture, came to embody both the sense of a fall from grace and also the bedrock, literally of all existence in the nineteenth century. The scientific investigation of fossil remains, combined with Darwin's theory of evolution, saw stone come to be associated with the interconnectness of all the planet's life-forms. Stone captured time and suggested continuity, reassuring people in that age of rapid change.

The Oxford Museum, with which Ruskin was involved, might be thought of as a 'proto-green' building, in that it tried to demonstrate the importance of Nature at this time of upheaval when the impact of industrialisation had left its mark on a 'green and pleasant land'. Surrounding the museum court are columns made from the full range of native stone and carved on their capitals are plants associated with the region from where they were quarried. But the building also embraced the (at that time) new material iron and has a glass roof. This material of the new railway stations was given the form of a Gothic cathedral, and the metal- and glass-covered museum was contained within a cloister-like space redolent of faith. The Oxford Museum shows how architecture can combine technical progress with more deeply rooted human needs expressed in everyday practice and made manifest in built form.

Futures

The future is likely to be exceptionally eclectic. Design will incorporate, for example, functional considerations that reflect regional climates as well as cultural preferences, local vegetation as well as high-technology materials from around the world. Some areas of a building's walls might well be in recycled bricks and others will reflect an ever-changing state of the art.

Eclecticism marked the end of the nineteenth century as architects tried to extricate architecture from style revivalism. It is to be hoped that our eclectic future will be more deep-rooted than this. Technical and directly environmental matters of sustainability will need to be set alongside the philosophical concerns with authenticity bound up in

Frampton's notion of a Critical Regionalism to resist the global exploitation of energy and material resources.[36]

The materials covering the surface will change. New, more efficient glasses, including vacuum glazing and shading systems, will let more light in while reducing the heat gain when it is not beneficial. The development of translucent aerogels introduces the possibility of combining high levels of insulation in a wall that allows the passage of diffused light. Photovoltaic materials will be used more and more to clad our buildings which will be orientated and shaped (or 'solar sculpted') to capture solar energy for light, heat and electricity. Intelligent skins that change according to the light (as special Polaroid glasses do) or temperature conditions will become widespread. The interior will also change. We will see surfaces that can change their colours and their thermal properties. Walls, for example, will incorporate phase-change storage materials to prevent overheating during summer days (see Glossary). Generally, we will see more materials that sense their environment and react accordingly, just as biological organisms do, and occupant control of conditions will become easier and more effective. We can also expect the further development of roofs that combine solar collection, daylighting (through rooflights) and rainwater collection – all these things happening now to become commonplace and more sophisticated. As buildings become more airtight the control of moisture movement becomes more important and we will have surfaces that can store and then release moisture.

The form of the buildings themselves will also change for other reasons. One is that the natural forces of the wind and sun are less powerful, or more precisely, have lower power densities (see Appendix A) than the mechanical and electrical systems that have powered our buildings over the past decades. (For example, the sun's radiation entering through a window at noon in October might be about 500 W/m^2 whereas the heat output from a radiator is about 1,000 W/m^2.) Thus, in order to use these energy sources fully, more surface area is required and this will inevitably have an impact on a building's shape and form. And because the surface needs to work with the interior, we can also anticipate changes in the internal spatial arrangements, in the heating, in assisted natural ventilation systems and materials, including the phase-change storage materials mentioned above. On these future forms we shall see also attached some unusual structures such as wind turbines of all sizes. The challenge will be to see if they can be made as elegant as a peacock's feathers. These trends are discussed in more detail in the chapters that follow.

Mies van der Rohe used to describe his concrete frames with their glazing as 'skin and bones' architecture. That particular combination of skin and bones depended on almost unlimited quantities of fossil fuels. What we are in the process of seeing is an incredible change in the 'skin' particularly, with the 'bones' to follow, we believe. This will involve more complex structures and materials that will be closer to the multifunctional skins of living organisms. To take but one example, in dry climates leaves adapt in a variety of ways including silvery shades to reflect light, fine hairs to trap moisture and waxy coatings to reduce water loss.

These are fundamental changes – from the passive to the active – that will influence our conception of buildings and cities. As we have seen, sites and building design are inextricably linked. And as we live in a more and more urbanised world which is beginning to realise the potential of, and need for, ambient energy sources we can expect to see more energy generated closer to the site. Thus, we are likely to see wind

4.24 Photomontages showing how wind turbines could more suitably be located in retail car parks, motorways and dockyards than on hilltops. Photomontage by Timothy Garnham.

turbines dotted throughout cities – on buildings, in car parks and along motorways and river-banks (Fig. 4.24).

At the end of the day, there will not be one fixed, correct way to make a sustainable architecture, but rather as many as we as individuals can create; hopefully informed individuals using sound environmental strategies and an inclusive, coherent approach to the needs and aspirations of human dwelling in the fullest sense. In his essay *Englands of the Mind*, the poet Seamus Heaney shows how Ted Hughes, Geoffrey Hill and Philip Larkin, all twentieth-century poets, construct a poetic language that evokes a different sense of place, but also of time. Hughes' is primeval, Hill's medieval and scholastic, Larkin's is more the modern urban (or suburban) man.[37] A sustainable architecture might be many and varied, ranging from a poetic expression of thermal mass in building to a use of more intelligent technologies and information systems in buildings making use of lightweight materials. In a similar way to that described by Heaney, we can expect architectural 'languages' derived from individual poetic sensitivities, traditions, climatic responses, sites, forms, technologies and materials to give us a great diversity of sustainable designs.

Building references

Benenden School: Architects – van Heyningen and Haward; Environmental Engineers – Max Fordham; Structural Engineers – Price and Myers.

Heelis National Trust Offices: Architects – Feilden Clegg Bradley; Environmental Engineers – Max Fordham; Structural Engineers – Adams Kara Taylor.

Portcullis House: Architects – Michael Hopkins and Partners; Environmental Engineers – Ove Arup and Partners; Structural Engineers – Ove Arup and Partners.

School of Slavonic and East European Studies: Architects – Short and Associates; Environmental Engineers – Environmental Design Partnership; Structural Engineers – Martin Stockley Associates.

BRE Building: Architects – Feilden Clegg Bradley; Environmental Engineers – Max Fordham; Structural Engineers – Buro Happold.

5

Heating, cooling and power

Introduction

In most climates, building design alone can not provide the degree of comfort that we need and, so, the development of culture and civilisation have been marked by our ability to heat and (less commonly) to cool the spaces that we create. This exceptional inventiveness is an ongoing characteristic of man but adapting to the environment is a requirement for all life and all living things must monitor and respond to their external environments. Throughout the animal and plant worlds one sees mechanisms for absorbing solar radiation or avoiding it, and for conserving heat or for losing it. The 'built' environment similarly modulates between exploiting solar energy or protecting against it, as we saw with *Homo erectus* and shelters built under cliffs (see Figure 2.2).

Besides adapting to the solar radiation environment, the natural world has also adapted to moisture and wind conditions. For example, when plants moved out of an aquatic environment onto land (about 500 million years ago), one of the key changes necessary was a way of reducing water loss. This was achieved by the aerial parts of the plant being covered with an impermeable, waxy cuticle (i.e. outer layer). This feature is with us today and is particularly notable in plants adapted to hot, dry summer climates. Examples you may know include cactus, box and viburnum.

Plants are of interest to designers from many points of view, including their ways of capturing light (more plentiful on land, of course, than in the seas) for photosynthesis. However, we have chosen to illustrate our main points more frequently with animals for a number of reasons. There is, of course, a certain self-interest (along with probably greater familiarity). And there is the fact that buildings, albeit stationary, are in some ways very close to homeothermic animals. (Homeotherms maintain a fairly constant body temperature by internal means; poikilotherms, by contrast, are organisms whose temperatures vary with their surroundings.)

Many plants and animals react to the sun – sunflowers move with the sun and orang-utans, as early as 14 million years ago, may have been using leafy branches to build sun covers (or to provide extra protection from rain) as they do today.[1]

The mythical origins of architecture as described by Vitruvius have man sheltering from the sun under branches. And as we saw previously according to him the idea of the primitive hut came from fallen branches over tree trunks. However, it is probably only with the appearance of man that the sun and heat, noticed most strikingly in the form of fire, start to take on a symbolic role.

This chapter looks very briefly at how the internal environment has been controlled and comfort provided through both heating and cooling in the past, speculates on how this might be done in the future and looks at the historical and design context of some of these functional considerations.

Physiology and heat transfer

Before considering how best to control temperature it's worthwhile taking a brief look at the heat transfer mechanisms of conduction, convection and radiation and at animal physiology.

Conduction involves the transfer of energy between adjacent molecules. (The thermal conductivity (see Appendix A) of air is lower than most solids and liquids and so trapped and stationary air is a good insulant – we will see how organisms and buildings use this further on.) Convection is the process by which heat is transferred by movement of a heated fluid such as air or water.[2] If we consider a hot surface and a cold fluid, the fluid in immediate contact with the surface is heated by conduction. It thus becomes less dense as it expands and rises, resulting in what are known as natural convection circulation currents. Convection that results from processes other than the variation of density with temperature is known as forced convection and includes the movement of air caused by fans. Radiation is the transfer of energy as electromagnetic radiation (see Appendix A) from one surface to another.

If we look at Figure 5.1 we can see a number of ways in which this fine

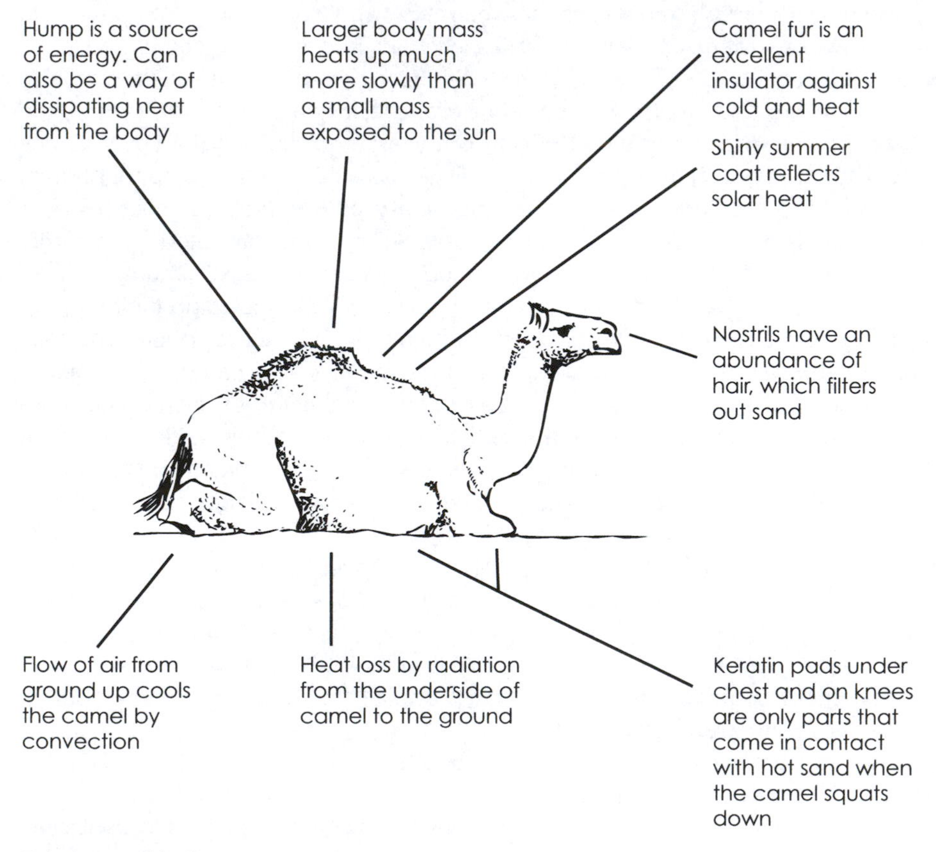

5.1 A camel in the sun.

homeothermic mammal, the camel,[3] maintaining its temperature at about 37°C (with an ability to let it vary from, say, 34°C to 41°C if required),[4] has adapted to its environment. When sitting to rest, for example, to protect itself from the radiant heat of the sun, and, so, to reduce the heat gain, the camel gathers itself into a compact form and faces the sun with the minimum surface area exposed to its rays.[5]

To guard itself against the hot desert sand the camel has keratin pads (keratin, a reasonably good insulator, is also what makes up your fingernails) under its chest and on its knees. These are the only parts that come in contact with the sand when the camel squats down. The ground underneath the animal remains relatively cooler than our delightful beast thus allowing for some heat loss by radiation. In addition, the flow of air up from the ground shaded by the camel cools by convection.

Conductive heat loss in animals is also commonly reduced by trapping air and so creating an insulating pocket or layer. In cold weather, birds do this by puffing up their feathers. The polar bear is protected against the Arctic winter cold by its dense fur and thick layer of fat which together form an insulating envelope. To limit heat loss, the bear's fur hairs rise, trapping a larger layer of air around the body and increasing the thermal insulation. The polar bear's hairy paws and ears and short tail all offer further protection against the cold.

In plants, bark often has a similar insulating role, providing protection from freezing winter temperatures.

The same physical phenomena of conduction, convection and radiation, of course, apply to animals and buildings alike. The building 'skin' – its walls, floor and roof – as we have seen need to protect against excessive heat loss or gains.

For example, in the igloo heat loss is reduced by using snow with its entrapped air and in addition by forming an inner lining with skins which traps more air (see Fig. 5.2). The use of snow as an insulator was, of course, not discovered by man.

Figure 5.3 shows how our friendly polar bear, having dug a den in the snow on the hillside, spends a quiet winter. During this period the bear's metabolism slows significantly and its temperature falls somewhat, thus, reducing its heat loss and providing a seasonal counterpart to the camel's daily variation in body temperature. In Shakespeare's home and many others of that period additional insulation was achieved by lining the walls with decorated linen sheets rather than skins (see Fig. 5.4) to trap air. This theme of insulation and decoration is a common and recurrent one. At Sutton House, London, started in sixteenth-century Tudor England, linen fold wood linings increase the comfort of the principal spaces (see Fig. 5.5). Their appearance is an 'ennoblement' of earlier fabric wall-hangings.

Convection and radiation

Figure 5.6 shows how human beings lose heat. Very approximately, a lightly clothed person in a heated home might lose 25 per cent through the skin and lungs and less than 1 per cent by conduction to the floor. The rest is lost by radiation to the surrounding surfaces and by convection (free convection is the main convection mechanism in rooms with low air speeds). The ratio of radiation to convection will vary with the precise conditions; for our lightly clothed person the amount lost by radiation is greater than that lost by convection. (See below for a comparison with a building.)

If we go back 250 million years or so to the Permian period and examine the reptile, the dimetrodon, one of our direct ancestors, we note the extraordinary sails on

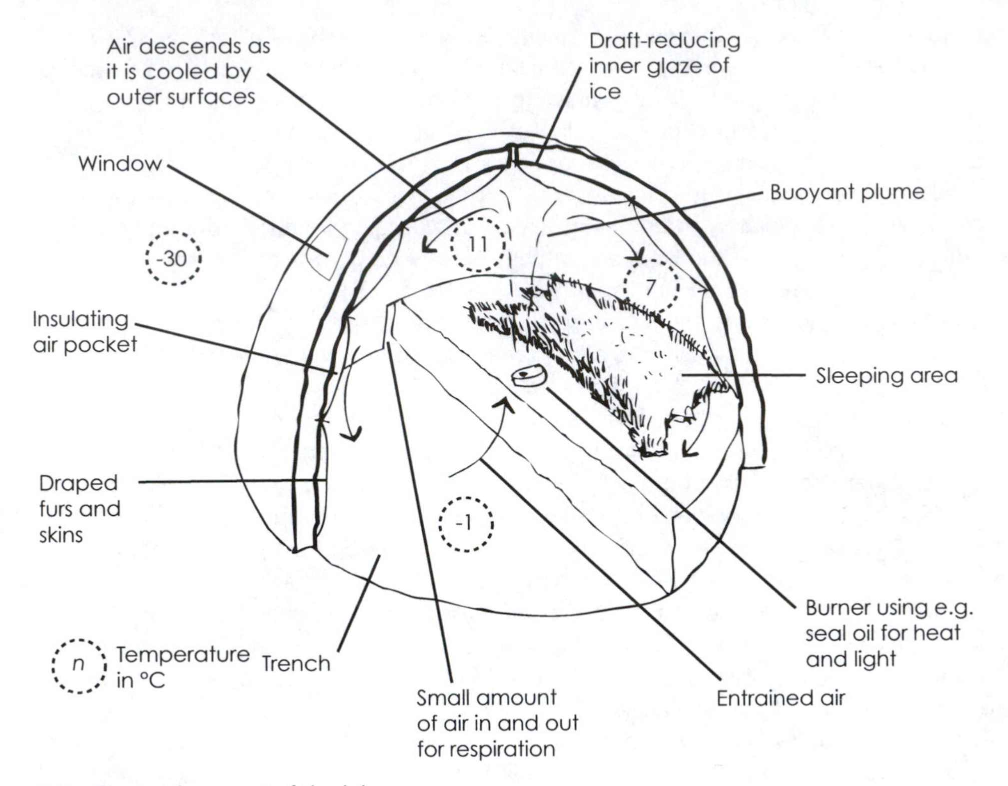

5.2a The environment of the igloo.

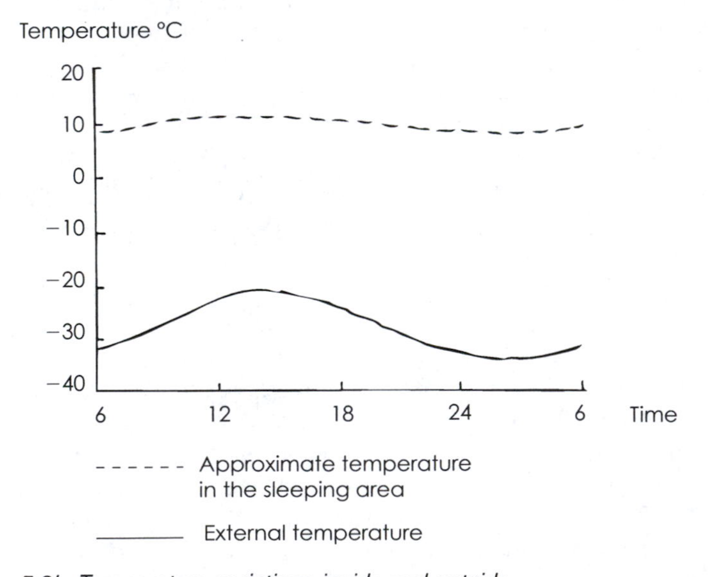

5.2b Temperature variations inside and outside.

Notes

a The sketches and data are based on a variety of sources including notes 7 and 8.

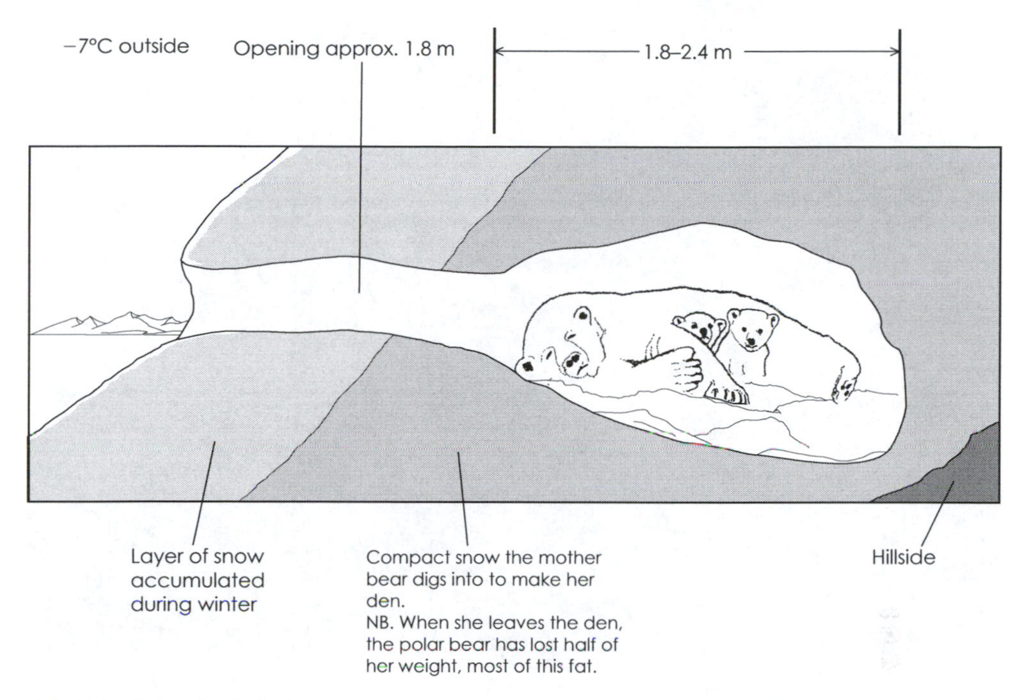

5.3 Polar bears in their den.

5.4 Shakespeare's home.

its back (Fig. 5.7). One of their possible roles is thought to have been in controlling body temperature.[6] By circulating more blood to the sails the dimetrodon could lose more heat both by radiation and convection. So, for example, in a strong wind this endearing reptile would lose more heat just as a building will do. The fins on the cylinder of a motorcycle engine work in the same way, enlarging enormously the surface area available to be cooled by wind passing across it.

5.5 Sutton House.

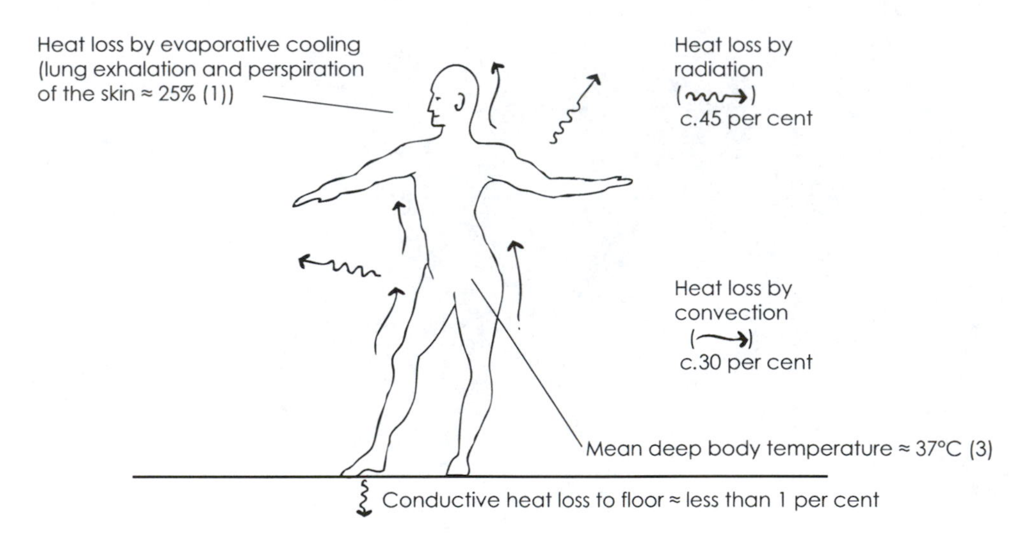

Total heat loss for a standing human adult – roughly 100 W (2) (Male 1.8 m² skin surface area, 104 W; female 1.6 m² skin surface area, 93 W (4) figures for a lightly clothed male in a room at 22°C. (Note that output is about 55 W/m² of surface area.)

5.6 Heat loss in humans.

1 Note 9.
2 Note 10.
3 Notes 11 and 12.
4 Note 13.

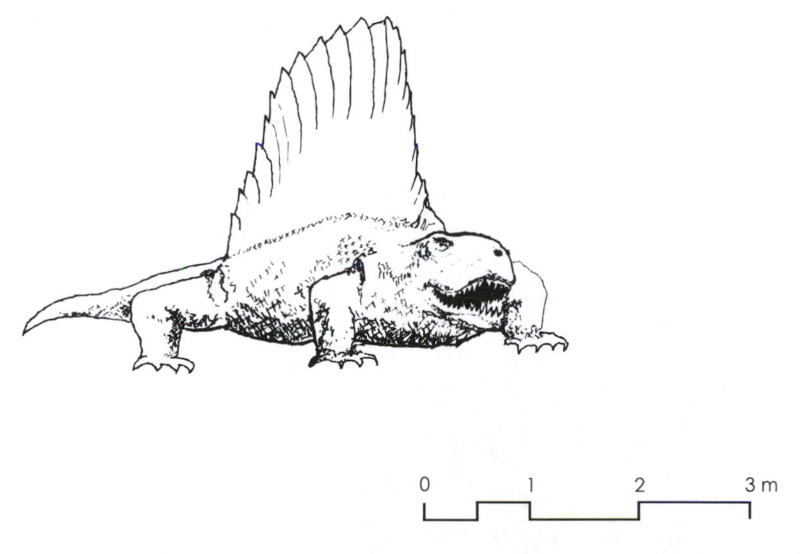

5.7 Dimetrodon.

Some 50 million years or so after the dimetrodon during the Jurassic period, stegosauruses (see Fig. 5.8) appear with a similar form also believed to be involved in temperature control. (Another hypothesis that has been advanced is that the scales are a sexual display structure.)[14]

And to return to the present once again, a wonderful illustration of convective heat transfer is the image shown in Figure 5.9 (using Schlieren photography – see Glossary) of cool air falling from the bottom of a frozen pizza.

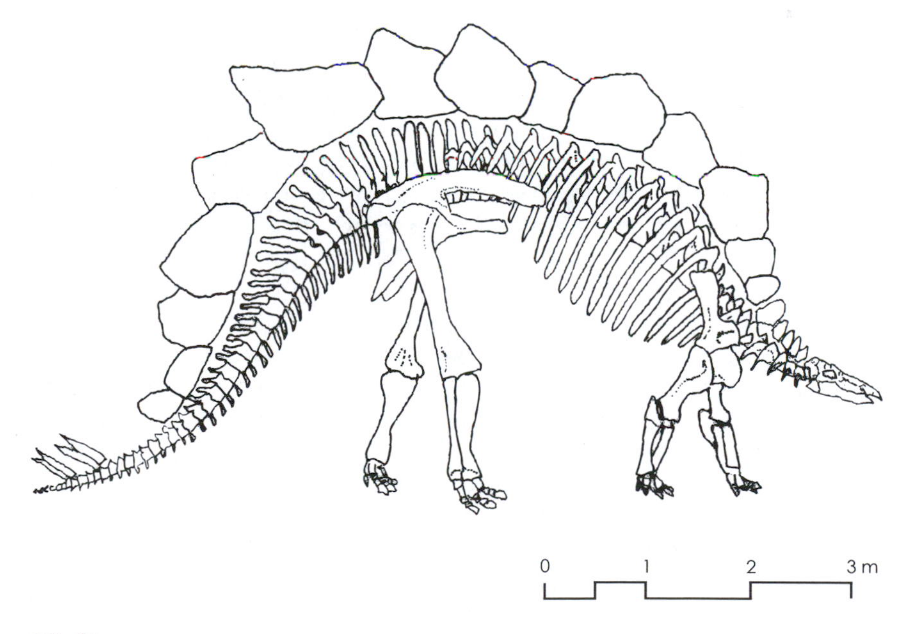

5.8 Stegosaurus.

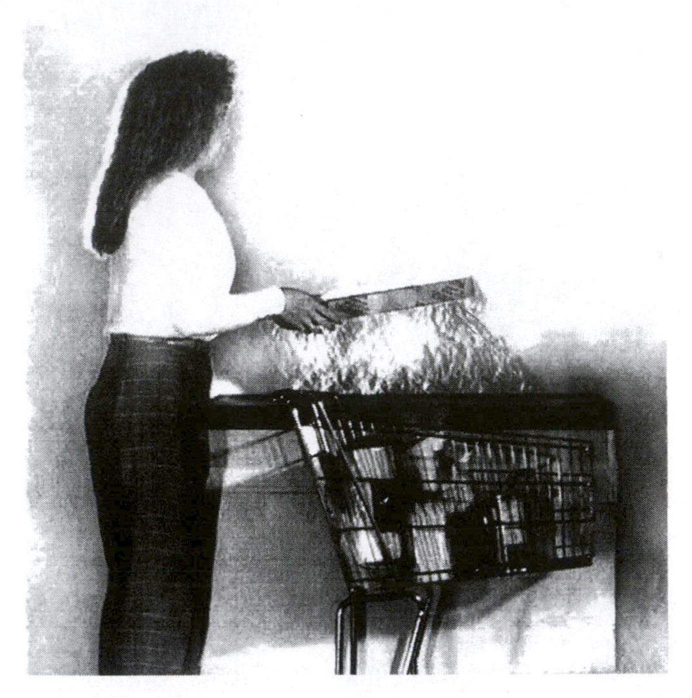

5.9 Cool air falling from a frozen pizza.[15]

A somewhat similar phenomenon occurs in the igloo (see Fig. 5.2a) where the coldest air falls by convection into the 'pit' or trench on the left side of the drawing.

Buildings

In buildings the heat loss has two components – the fabric heat loss and the ventilation heat loss (roughly comparable to the heat loss via the lungs in mammals).

A measure of the fabric heat loss is the U-value which takes into account all three mechanisms of heat transfer: conduction, convection and radiation – the lower the U-value, the better the insulation. (See Appendix A for a more detailed discussion.) And, of course, the better the insulation the less energy needed to maintain comfort and so, in principle, the lower the environmental impact. For a rough comparison with a human being, on a clear day in February in London a modern well-insulated[16] and well-sealed house might lose approximately 30 per cent of its heat by ventilation, 50 per cent by convection and conduction and 20 per cent by radiation.

Ventilation heat losses can be reduced by careful detailing of the building during the design stage and by meticulous construction to ensure that air enters only where and when required. By controlling the heat losses and gains of the building skin and by controlling the ventilation movements through it, the energy requirements for heating and cooling can be reduced. Doing this increases the practicality and economic viability of renewable energy sources discussed below.

Heating

In some lands which are blessed the sun alone provides all the heat required for homes. However, as we have seen earlier in other climatic regions, orientation, massing and the need to provide heat for spaces, water and cooking have an important influence on the form of settlements and buildings.

The Magdalenian people who produced the cave paintings in Lascaux in France approximately 17,000 years ago lived in tepees (Fig. 5.10), as did many native American tribes until quite recently. The heat loss of such a structure on a winter's day might have been about 1.5 kW, and so a small fire with wood as a fuel would have been sufficient for comfort. Note that comfort is maintained in the igloo (see Fig. 5.2a), with the heat of an even less powerful oil burning lamp with an output of, say, 0.3 kW. (The lamp was also used to provide heat for cooking. Because energy was scarce about half of the Inuit diet was eaten raw.)

A centrally located fire, as mentioned in Chapter 2, would provide a focal point for warmth and conversation and engender a sense of security. Our cultural memory of this is probably quite deep.

At the centre of Skara Brae houses, there was a fire which probably burned dried seaweed or dried animal dung – there were very few trees on Orkney and peat began to be formed only after *c.*3000 BC when the climate deteriorated. (Recent archaeological research has established that milk and butter formed part of the inhabitants' diet and perhaps the small side chambers provided a cooler space for storing these – an early larder in fact.)

From such humble beginnings, the role of the fireplace grew with both technological and cultural innovations. By the sixteenth century in Tudor England elaborate

5.10 Magdalenian tepee.

and decorative chimney-stacks were being used by the wealthy as a means of display.[17] (At the Contact Theatre which is discussed in the next chapter tall ventilation stacks signal the presence – and symbolically the aspirations – of the theatre which is somewhat removed from the main road.)

There were experiments with stoves in England in the seventeenth century but it was later that real progress was made. Around 1740 in America the inventor Benjamin Franklin turned his attention to how one could extract more useful heat from wood as a fuel. He developed the eponymous stove which could be installed in a fireplace and because of its compact form and design (including a convection chamber) was considerably more efficient than an open fire and earlier stoves.

This is one example in an ongoing process of improving the efficiency of fuel use, of substituting one fuel for another and of moving towards fuels with less (and ultimately, no) carbon in them.

Domestic fires traditionally burned wood (when available). A shift to coal[18] occurred in the eighteenth and nineteenth centuries. When readily available supplies of wood, coal and whale oil began to diminish, oil exploration was stimulated. The first well drilled for oil was in Pennsylvania in 1859. The first important modern use was as a fuel to replace whale oil in lamps. Subsequently petroleum was used in transportation and for heating. As we saw in the last chapter in less than 50 years oil, in conjunction with a number of other key inventions such as the electric motor, transformed the world's economy and its architecture.

However, varying technologies advance in a rather disordered fashion. In the early part of the twentieth century coal was still being used in domestic boilers (the first domestic oil burners appeared on the US market in about 1920).[19] But advances in pumps, boilers and central heating were having a major effect on architecture. In this context Frank Lloyd Wright was able to articulate the forms of his Prairie houses for light and air and for making 'a house in a garden or the country the delightful thing in relation to either or both that imagination would have it'.[20]

Rooms were often heated by radiators[21] – something of a misnomer since about 65 per cent of the heat is given off by convection and only 35 per cent by radiation. One device Wright used was to place radiators under window seats which were slatted to allow the warmed air to circulate by convection.

The skyscrapers of Mies van der Rohe, and those who followed him (see Figure 4.7) were inconceivable without oil for heating and oil for power stations to produce electricity. With time the use of natural gas increased as it became another source of (fossil) fuels for our energy-guzzling cities.

Part of the legacy of this, though, is that the ready availability of energy for many during the twentieth century appears to have prevented more thought being given to more efficient forms and 'skins' for buildings. In ecological terms, there was no pressure for a better adaptation of the 'organism' (i.e. the building) to its environment. (And so the building fabric with its relatively poor insulation levels changed very little for decades in the middle part of the twentieth century.) Indeed, the burgeoning oil industry in the American south-west in the 1930s wiped out the fledgling solar water-heating industry that produced the Los Angeles solar collectors of 1900 shown in Figure 5.11.

In mid-century, transparency and oil (an odd couple in any other context) dominated with huge areas of glass characterising office construction and poorly insulated envelopes found everywhere in the domestic sector. But the approach to design could

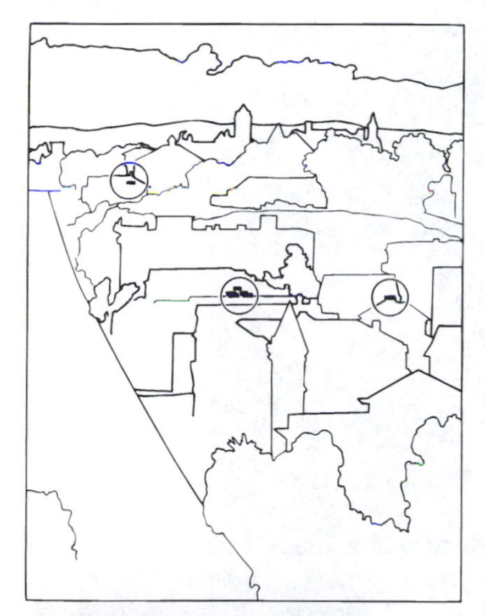

5.11 Los Angeles solar collectors.[22]

be quite eclectic. If we look at Mies van der Rohe's renowned 1951 Farnsworth House (see Fig. 5.12a), we see why its elegant lines and cool steel and glass construction made it the epitome of modernity. And how is it heated? In part, by a vestigial device, well known to our ancestors, a place of fire, or more currently, a fireplace, carefully tucked away into the services core (see Fig. 5.12b).

Although solar houses were being developed in the 1950s and 1960s it was really only in the 1970s that in response to the 'oil crisis' an environmental movement began that has gradually gained momentum, unfortunately due to the problem of global warming.

In the 1980s and 1990s a number of buildings started to seriously address general concern. The BRE Environmental Building (1993) introduced in the previous chapter (Fig. 4.23) was an important step towards an architecture of low demand with supply being provided at least in part by renewable energy sources. Use of passive solar gain, thermal mass, high insulation levels, natural ventilation, daylighting, borehole cooling and photovoltaic panels all contribute to creating a comfortable environment with a much reduced demand on the environment.[23] Highly efficient heating and cooling systems along with the ventilation are run by a building management system which receives data from external and internal sensors (similar

a

b

5.12 Farnsworth House.
a View.
b Fireplace.

to the functioning of an animal's brain). Similar principles were used at Kings Langley discussed below.

Cooling

In the animal world, a prime way of maintaining a cool body temperature is to move towards the shade and, so, for example, in the desert certain lizards and snakes bury themselves in the sand which acts as a shield. Obviously, it is more difficult for buildings to do this and so they have tended to stay in place and rely on either the diurnal temperature range (see below) or local winds or both to help maintain temperature. Compact forms which expose less surface area to the sun (similar to a camel at rest) and thermal mass to absorb heat during the day and lose it at night are also important.

In buildings, cooling and ventilation are often linked. This can be done above ground – the BRE Environmental Building, for example, is largely cooled in this way. Here advantage in taken of the diurnal temperature range. With the setting of the sun, air temperature tends to drop. The difference, over 24 hours between the maximum day temperature and the minimum night temperature is the diurnal range. In the summer in England this is approximately 10°C (say a 27°C peak in the afternoon and a 17°C minimum in the early hours of the morning).

It is also possible to link the building with the ground temperature. Below-ground temperatures are lower than surface temperatures in summer (see Appendix A). In the sixteenth century Palladio wrote about the Costozza villas (Fig. 5.13) which were built above natural caves and used the cool air to ventilate the house via marble grilles.

Palladio was also sensitive to the environment in his own work. His Villa Rotunda, located on the crown of a small hill outside Vicenza, expressed the house as temple and its iconic power had enormous influence, copies being made all over the world – including by Jefferson at Monticello in the USA. Palladio's 'Four Books on Architecture' was a typical Renaissance treatise combining poetics with the most pragmatic matters of building. It was the harmonious order, symmetry and the temple porticoes that caught the eye, but Palladio described the porticoes in purely pragmatic terms as being designed to receive the breeze and gain the view of the 'four quarters'. The building also very importantly draws cool air from the basement below and distributes it to the spaces above.[24]

The Matmata dwellings in the Sahara, as we saw in Chapter 2, are built into the earth. By careful attention to form (the relations of width, breadth and depth) to limit the sun's penetration into the courtyard and by use of thermal mass the effect of the sun is tempered. Whereas the peak daytime air temperature is 45°C, the peak inside is only 28°C; the diurnal variation in the Sahara can be over 20°C (see Fig. 2.5).

Extreme heat has always prompted ingenious responses. At the Amber Palace in India, constructed over a period of about 200 years starting in 1592, a wall in one room of the harem had water from a mountain stream flow over it; to enhance the sense of repose the water was perfumed.[25]

On a more prosaic note, in the late nineteenth century one of the first buildings to mark the tentative beginnings of air conditioning (i.e. the control of temperature, humidity and composition of the air) was the Royal Victoria Hospital, Belfast which had coconut fibre ropes that were wetted in the incoming air stream to lower the temperature and filter the air.[26] Wind catchers in Iran often had a palm frond mat soaked in water placed at their opening.

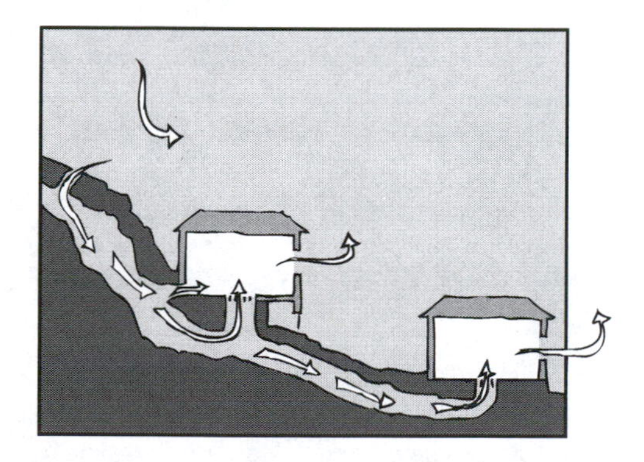

5.13 Costozza villas.

In current construction where additional cooling is required, there are a number of other solutions ranging from using water in aquifers to absorption chiller units (see Glossary) which use solar energy. At Snape (see Chapters 2 and 4 and cover photo), for example, water at about 13–14°C is drawn from an aquifer at a depth of about 5 m and used to cool the air to the concert hall and as we have seen at the BRE Environmental Building water at 12°C is drawn from a depth of about 70 m.

A key issue for buildings is that somehow the heat generated inside them, by the occupants and the processes – lighting, computers, cooling and so on – needs to be taken away. Ventilation, as we have seen and as discussed in more detail in the next chapter, can often deal with this. In some cases, though, there is a need for a different approach using more complex systems to remove heat from an area of high heat production and discharge it to the external environment. Traditional mechanical cooling systems do this – for example Figure 5.14a shows schematically and in a very simplified way how at the Kingston University Faculty of Design heat is removed from TV studios deep within the building and via the intermediary of a chiller is discharged to the outside air passing over the roof by dry-air coolers (see Glossary).

Figure 5.14b shows a somewhat analogous process.[27] In dolphins the testes are deep within the body but need to be cooled so that sperm production is maintained. This is achieved by the arterial blood supply to the testes passing through a heat exchanger where it is cooled by blood in veins which has come from the fin exposed to the cool water of the sea. (A similar heat exchange system provides cooling to the female dolphin's uterus to maintain the correct temperature for the foetus during pregnancy.) The dolphin fin is thus a marine equivalent of the dimetrodon's scales.

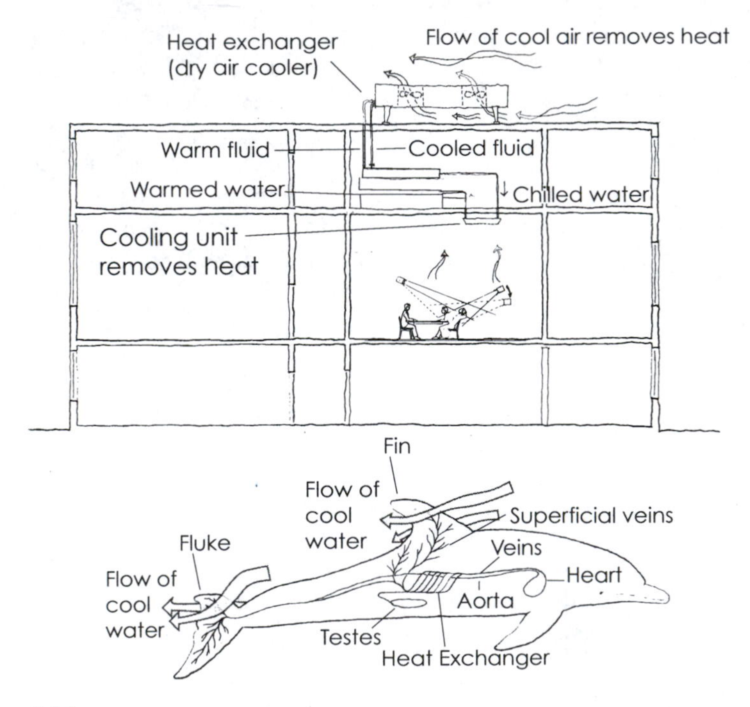

5.14

a Heat removal at the Kingston University Faculty of Design (simplified).

b Heat removal in a dolphin.

An advantage of systems that rely in whole or in part on liquid coolants rather than air is that the higher specific heat capacity and densities mean much greater quantities of heat can be transferred (1 m^3 of water absorbs 4,000 times as much heat as a cubic metre of air for, say, a 10°C temperature change). Blood which, of course, is mainly water, is a very good medium for heat transfer.

Power

Electricity in buildings is a relatively new phenomenon – it first appeared in about the 1870s in the homes of some wealthy enthusiasts and also in some more far-sighted public and private communities where it was used mainly to replace or supplement gas lighting. Very quickly it was adopted in a number of public and industrial buildings and then, as mentioned previously, spread to homes for lighting and power to run the appliances which reduced the amount of manual work that had to be done.

At the outset distribution networks were, of course, quite limited and varied from individual country houses to villages. In part, because of this combined heat and power (CHP), installations which supply both thermal energy and electricity and are often more efficient thermodynamically were economically viable. For example, the American inventor Edison had companies in Albany and Boston providing both heat and electricity in 1887.[28] In the first half of the twentieth century, though, the development of large-scale grids led to the demise of CHP. However, we are likely to see a growth in its popularity in the next 20 years with technological advances at both domestic level, for example, fuel cells (see Glossary) and urban level, for example, heat recovery from waste. Such a development would be entirely in keeping with our philosophy of the democratic and efficient use of local resources.

The past century has seen people spending more and more of their time in buildings, higher densities of occupation and increasing electrical loads. Power would appear to have become essential for lighting, for pumps and motors for heating, cooling and ventilation, for lifts and for everything from electric toothbrushes to computers.

The fossil fuel and nuclear power stations that have met the bulk of this load have had an enormous environmental impact and the challenge at hand is to replace them with more sustainable approaches.

As was mentioned in the previous chapter we can expect to see more energy generated on and near the sites of our buildings. Both wind turbines (see the following section) and photovoltaic power will produce a growing amount of electricity for us in the future and both will have an important impact on building design.

Figure 5.15 shows the housing scheme at Parkmount in Belfast which was designed especially for photovoltaics on the roof. (These roofs may be viewed as a series of abstract 'leaves' arranged to capture the maximum amount of the sun's radiation). Photovoltaics have been installed on the roof of the bottom of Figure 5.15b and the intention is to install them on the other roofs in the future.

Present trends and the near future

The present is characterised by two significant trends – the climate and our natural environments are changing for the worse and we are thinking more about the consequences of this and what needs to be done.

We can expect more variability in the weather with greater incidences of extreme

a

b

5.15 Parkmount housing.

conditions and in Europe higher summertime temperatures. This will mean that building designs will need to consider even more carefully their heating and power demands and especially how to minimise artificial cooling requirements.

Better construction and tighter envelopes ('skins') should reduce energy demands and cooling loads. Improved and new technologies will reduce loads (more efficient lighting, computers, power transmission cables and so forth) and provide new ways of supplying demand. As demand is reduced it becomes more cost effective to provide some of the supply on site.

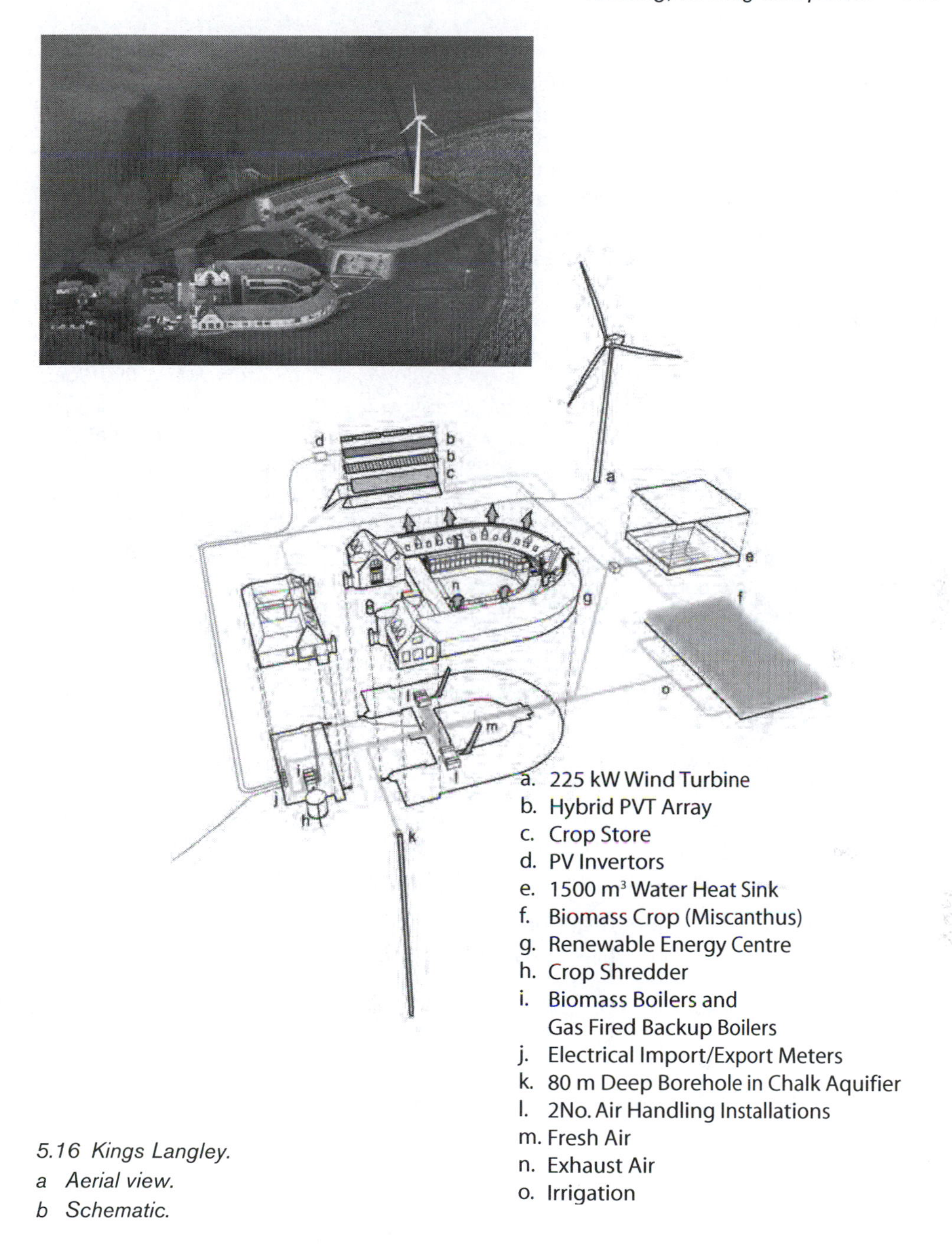

5.16 Kings Langley.
a Aerial view.
b Schematic.

Holistic integration of the site, building, energy supply and water and waste systems will become more common as we suggested in Chapter 2. A striking example of this is the Kings Langley project by Studio E Architects shown in an aerial view and schematically in Figure 5.16. This is the headquarters of Renewable Energy Systems who develop wind farms.

Here demand has been reduced by high insulation levels and energy-efficient equipment. The full potential of the site is exploited by a 29 m diameter wind turbine which provides electricity to the buildings and the national grid, by growing miscanthus (a grass) which is dried and fed to a biomass boiler for space and hot-water heating. Solar energy is used for electricity production and water heating. Hot water is stored in a 1,500 m^3 tank in the ground insulated with a lid. Cooling is provided from an 80 m-deep borehole in the chalk aquifer.

Other developments in energy supply which we can expect to see are combined heat and power (CHP) plants. As mentioned above, these have been in use for some time but new units are much smaller and can operate at the level of the home. Energy from waste is also an area that will burgeon.

A logical and possible – many (including the present authors) would say necessary – development from projects such as Kings Langley is to move towards a hydrogen economy.[29] This, in its most 'environmental' form, involves the production of hydrogen using renewable energy sources such as wind turbines and distribution on a national (and, probably, international) scale to supply our transport systems and buildings either directly or in fuel cells. These devices combine hydrogen with oxygen from the air to provide electricity, heat and water! Although this may sound like a modern version of the alchemist's dream, it is not and we can expect to see the first serious steps towards implementing it in the next decade. If successful, it will give us a significant basis for a more sustainable society that can help restore our natural environments.

Further reading

Givoni, B. (1994) *Passive and Low Energy Cooling of Buildings*, New York: van Nostrand Reinhold.

Thomas, R. (ed.) (2001) *Photovoltaics and Architecture*, London: E. & F.N. Spon.

Thomas, R. (ed.) (2003) *Sustainable Urban Design*, London: E. & F.N. Spon.

Thomas, R. (ed.) (2005) *Environmental Design*, London: E. & F.N. Spon.

Building Services Engineering cited in note 19 (p. 217) is indispensable for anyone interested in the historic development of heating, cooling, ventilation, electricity, etc.

Building references

Parkmount Housing: Architect – Richard Partington Architects; Environmental Engineers – Max Fordham LLP.

Kings Langley: Architect – Studio E; Environmental Engineers – Max Fordham LLP.

6

Ventilation

Introduction

The Po Valley, the home of Parma ham, is surrounded on nearly all sides by mountains and when moist air moves across them it falls as rain. By the time the winds blow down into the valley it is much drier, resulting in ideal curing conditions.[1] The local population has appreciated this for centuries and so the drying racks for the hams are aligned in an east–west orientation to harness the maximum potential of the wind. Thus, a combination of the right site, environmental understanding and appropriate technology leads to culinary delight (at least if you're not a vegetarian or a pig).

A similar understanding of the environment is manifest in the area around the Arabian Gulf, the Gulf of Oman and the Arabian Sea with their windcatchers, referred to previously, and we will return to these devices below. What is particularly of interest is that these starting points are regional. They respond to the general climate of the area and then develop in the more particular context of a specific site or grouping of urban buildings. And also, of course, they are natural systems relying on renewable energy from the sun and wind.

Biological design has similarities and we will look at a number of strategies in a bit more detail. After that we will elaborate on some of the references made earlier to principles of ventilation. This will involve tracing very briefly the development of ventilation strategies in buildings and examining how ventilation in connection with heating and cooling, as appropriate, has contributed to the effective functioning of the building (i.e. the organism) and so to the comfort of its inhabitants. The challenge of a sustainable architecture is, of course, to develop a ventilation strategy which requires the minimum amount of energy while contributing to the form and appearance of the building. Both ventilation and light (as introduced earlier and as we will see in more detail in the following chapter) play a part in the generation of form in buildings, cities and living organisms.

A biological environment

The first primitive forms of life had, of course, to deal with heat transfer and radiation (just as our camel of Fig. 5.1) and find a way of maintaining their water balance with the environment. These forms were often unicellular and exchanges with their surroundings occurred at the envelope, i.e. the cell membrane.

As more complex life forms evolved, the need for respiratory ventilation became more and more important. Respiration systems were required both to provide oxygen so that energy could be derived from food and to take away the products of this metabolism. This is, of course, somewhat similar to buildings where ventilation brings in fresh air and takes away the waste products i.e. the stale air. A number of respiratory systems, for example, in insects, simply relied on oxygen being able to diffuse in from the outside. This is only suitable for small organisms and for life to grow larger as we saw in Chapter 3 more complex forms and systems are needed. This is the equivalent of developing the section at the same time as the façade. These new biological forms included such rich and complex respiration systems as that of our familiar camel (Fig. 5.1) which can filter air, humidify it, alter its temperature and control its flow rate (all with a greater degree of reliability than most of the world's air-conditioning systems).

One also sees a high degree of control and complexity in the constructions of some animals. For example, termite mounds (see Fig. 6.1) draw air in from the outside, taking it through a heat exchanger to maintain the air temperature at about 31°C even though the exterior temperature can reach 60°C.[2] By drawing water from the earth up into the structure it is also possible to maintain a constant relative humidity of about 90 per cent. The mound walls are about 50 cm thick and this provides both good insulation and high thermal inertia, similar to the case of the earth surrounding the sunken Matmata dwellings (Fig. 2.4).

Honeybees have a somewhat simpler system. The nest temperature needs to be between 32°C and 36°C for the eggs to develop and hatch.[3] If the temperature drops below this worker bees huddle together around the brood to keep it warm and if it gets too high they position themselves at the nest entrance and use their wings to fan out the warm air – true ventilators!

Plants

In many plants water is taken up by the roots and is lost as water vapour at the leaves in a straight-through process known as evapotranspiration. About 400 million years

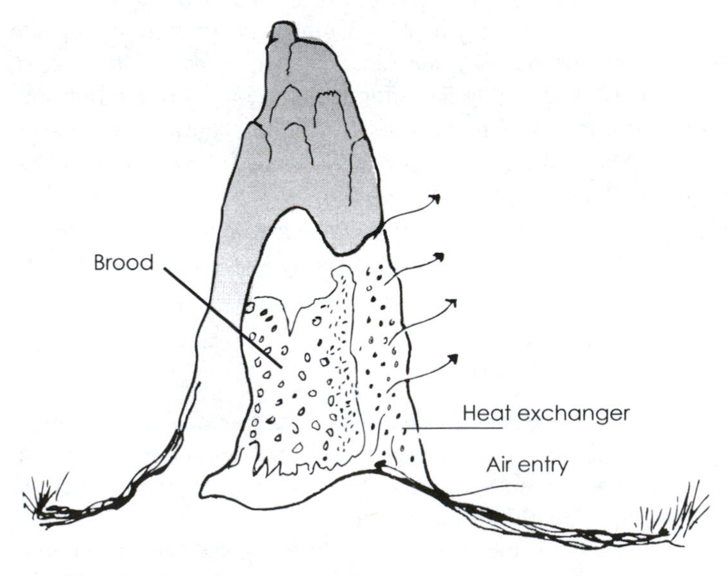

6.1 Termite mounds.

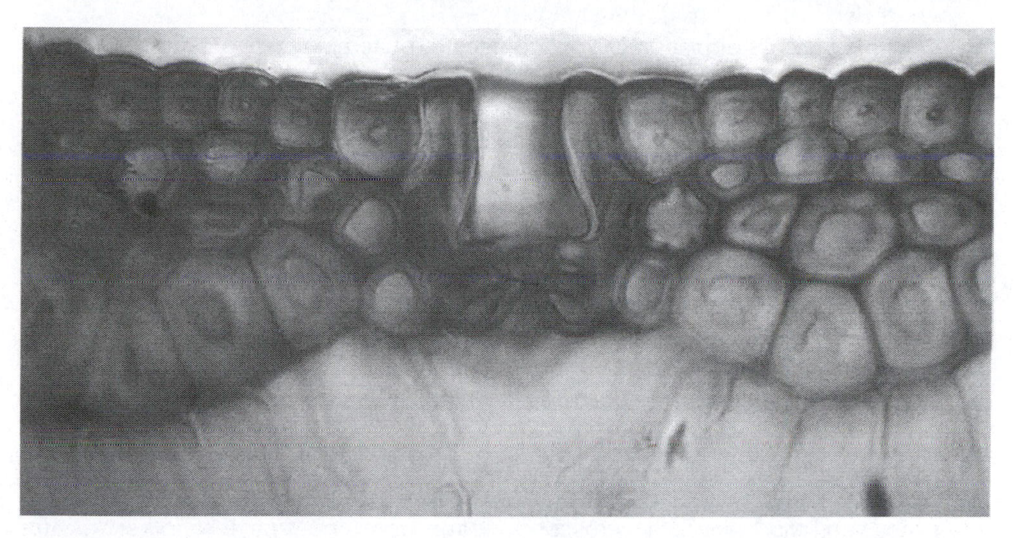

6.2 Stomata of a pine needle.[4]

ago plants developed stomata, structures on the leaf surface, to allow gaseous exchange – carbon dioxide for photosynthesis enters and oxygen produced in the process exits and during this water vapour escapes.

Figure 6.2 shows the stomata of a pine needle (magnified by a factor of about 4,000) – the stomata is located in the upper centre of the image on the outside of the leaf.

The stomata are thus modifications to the surface, allowing specialised activities to take place, similar to what we find in the animal skin or the building envelope.

Ventilation in buildings

Nature is a great equaliser. There is a tendency with time for things to become more disordered and for differences to become less distinct – steel rusts and the autumn leaves fall.

One aspect of this is that if two contrasting physical systems are put in proximity they will interact to become more similar – a sugar cube in a cup of coffee dissolves to produce a sweet liquid and an area of high pressure on the earth will spread out with time to reduce its pressure and interact with the lower pressure areas around it (see Appendix A).

The movement of air in buildings – known broadly as ventilation – of course follows these principles. Air moves from zones of high pressure to low pressure. As previously alluded to, ventilation is required to supply fresh air for breathing and to create a pleasant atmosphere; to remove pollutants (including CO_2 produced by us) and excess moisture to reduce the risk of condensation, and to take away a surplus of heat. As a comparison a human body in the open air inhales air with about 21 per cent oxygen and 0.04 per cent CO_2 and exhales air at about 15 per cent oxygen and 4 per cent CO_2 (air moves in and out of the lungs due to differences in pressure). Air entering a building will of course be similar to that entering the body – air leaving may be 20.5–21 per cent oxygen and 0.04–0.1 per cent CO_2. CO_2 levels in buildings are controlled either automatically or by people opening windows when it feels 'too stuffy'. (In environments such as airplanes CO_2 often rises to quite high levels of about 0.15 per cent; this is allowed because it reduces the need for fuel to heat up outside air at −50°C at an altitude of 10,000 m.)

In a mechanical system a fan creates an area of low pressure on one side to draw air in and a high pressure area on the other to distribute it to the building. Your lungs (in conjunction with the muscles of the chest and abdomen) use about 1 per cent of your energy intake and do the same thing for your body – for comparison, a fan in a building might create a pressure of 150 Pa and your lungs will vary from about −150 Pa to +150 Pa (above atmospheric pressure) as you inhale and exhale. As a reference, if you blow hard on your outstretched hand, the pressure on it might be about 0.05 Pa (see also Appendix A).

Natural ventilation systems in buildings have three main mechanisms (see Fig. 6.3):

1. Single-sided where air comes in from a window as, for example, in Melsetter House (Chapter 3);
2. Cross-ventilation where air crosses a space from one side to another (as, for example, in the Farnsworth House (Fig. 5.12));
3. Stack effect ventilation where heat loads inside a space cause air to be warmed and rise. Letting it out at high level and bringing in replacement air at low level can produce an effective ventilation system. (The De Montfort Queens Building and the Contact Theatre both discussed below are examples of this.) (A simple example found in single-sided ventilation is the common sash window first used in England in 1670 which lets fresh air in at the bottom and exhausts air at the top.)

These processes are 'natural' and are provided by pressure gradients (wind – as we are reminded when it blows a door shut) and temperature differentials (from heat given off) which in turn create pressure differences. They are 'straight-through' approaches – air comes in and then leaves. They have the great advantage of not requiring the large quantities of energy to run the powerful fans used in mechanically ventilated buildings and so are important for environmentally-friendly building of the future. However, because the driving forces are low, high-resistance elements such as filters must be avoided. Natural ventilation systems work best where the outside air is clean. A sustainable building requires a sustainable surrounding – and this in turn is only possible through a holistic approach that considers site and city as well.

6.3 Natural ventilation mechanisms.

Assisted natural ventilation systems are a step-up in terms of complexity. These incorporate low-energy fans to provide greater reliability and to cope with greater resistances than natural systems can deal with (see Appendix A). The key design issue is to strike the right balance between natural forces and fans but there is a second important question related to embodied energy. A low-resistance air path that requires huge amounts of space and large amounts of material is not necessarily an economical or environmental solution. (Living organisms, of course, face a similar challenge of finding the best way to organise a limited amount of physical material – a species that invests too much in mass and not enough in mobility may find itself an attractive prey.)

Buildings designed to work with natural forces are likely to have forms which are more clearly influenced by them than buildings dependent on mechanical power. One reason for this is that the energy densities and flows are lower in natural systems and so greater areas are required (see Appendix A).

Mechanical power, as we saw in Chapter 4, began to have a major effect on design in the second half of the nineteenth century. The development of electric motors and pumps allowed an explosive development in building services and in addition to fresh air it became easy to circulate heat, coolth and water around buildings. Electricity, of course, also facilitated a revolution in information technology in the form of the telegraph and telephone. (And in both buildings and cities pneumatic systems conveyed messages in capsules, a bit like the proverbial 'bottle'.)

These technological developments were not only used to develop tall buildings. Wright, for example, said that heating pumps and central heating allowed him to articulate the form of his houses.

A number of larger buildings started to rely on electricity for their fresh air supply. Mechanical ventilation, in contrast to natural ventilation, uses a system of fans and ducts (which can be formed from the building fabric or made of metal) to distribute outside air to the often deep interior spaces of a building. It tends to be energy-intensive but has an advantage of great flexibility. If you have enough energy you can deliver air to anywhere in a building. More complex mechanical systems can also recirculate part of the air in order to conserve heat or coolth.

Returning to natural systems, Figure 6.4 shows a typical courtyard home with a wind tower in Baghdad. Air enters the house and passes over a pool of water – the cooled air then is distributed to the living spaces by a system of openings to provide comfort. Exhaust air leaves via high level windows. A further elaboration is found in the lakeside city of Bam.[5] There the wind tower is separated from the house by a garden. Air caught by the wind tower is directed into a subterranean passage where it is cooled by the lower ground temperature, due in part to irrigation water seeping down. This structure ranges from the earth to the sky both actually and poetically, and admirably uses all the resources available to it.

Figure 6.4 also shows the wind catchers of houses in the city of Hyderabad in the Sind province of Pakistan – one of the finest architectural examples of this ventilation strategy. Orientated towards the prevailing wind they create a dramatic skyline and in this case a visual delight.

All ventilation solutions, whether vernacular or in our modern buildings, whether 'natural' or mechanical, need to respond to a variety of constraints, both environmental and cultural. In the igloo fresh air needs to be brought in for the occupants and the burner, but because it is extremely cold the air supply needs to be limited. In wind

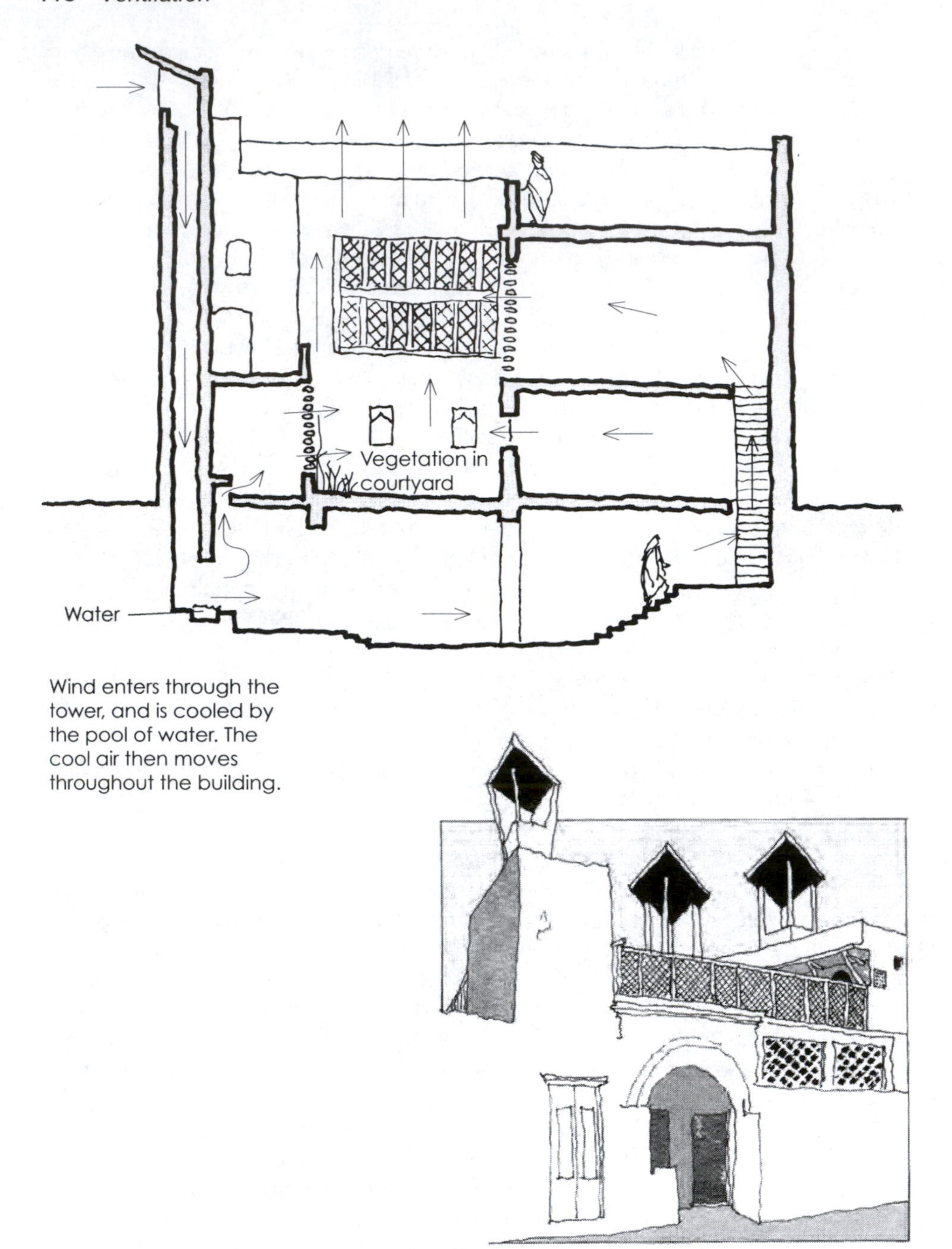

6.4 Wind towers.

towers we see a fine example of an architecture responding to wind, temperature and even water and coming to symbolise a culture.

Historically, shelter started with a simple approach to ventilation and continued like that for some time. Single-'cell' solutions were common. Where conditions were extreme as, say, yurts in the steppes of central Asia with winter temperatures of

−30°C, great care was taken in the selection of materials and the details of the construction and covering, i.e. the 'skin', to reduce heat loss. In more temperate climates less attention was given to ventilation and homes were often draughty.

The Acheulian hut (see Fig. 2.2) had a single opening which allowed oxygen in and CO_2 out. This opening might be viewed as a stomata performing the reverse of photosynthesis, of course, and a reminder of our dependence on the plant world. In the Magdalenian Period the tepee like structures let air in from an open skin (literally) flap and out through the top (see Fig. 5.10).

In the Iron Age in Britain (approximately 700 BC to AD 43[6]) huts (Fig. 6.5) had an entrance for movement and air, but where there was a hearth there was unlikely to be any special opening in the thatched roof. Air and smoke filtered through the thatch and, although this must have been quite uncomfortable for those inside the hut, it was clearly preferable to the thatch being set ablaze by a strong air current fanning a spark that had escaped from the hearth.

Later, houses often had holes in the roof covering to vent the smoke. This may have been the case, for example, at Skara Brae where the roof was probably made of turf on birch bark split open and laid like tiles. The smoke hole tradition was continued until quite recently in Orkney crofts. A farm steading at Kirbister (now Kirbister Farm Museum) is a good example of this. Here the system has been made more efficient by placing a T-shaped piece of wood (see Fig. 6.6) along one top edge of the square smoke hole. This leads to an acceleration of the wind and a reduced air pressure in the hole. The smoke is thus drawn out of the hole and carried away by the wind; in the process downdraughts are reduced. A handle attached to the piece of wood allows it to be moved to the edge of the smoke hole facing the wind when its direction changes. The effect of the wood at the leading edge is similar to that of aerodynamic lift in an airplane where the wing shape leads to the creation of zones of high and low pressure .

A similar but updated solution has been used by the architects Jesticoe and Whiles in their transformation of a Covent Garden warehouse into offices – incidentally another good example of the sustainable reuse of structures. In that building an existing lightwell was transformed into an atrium which serves for both light and ventilation. At the top of the atrium is a skylight (see Fig. 6.7) with the profile of an airplane wing

6.5 Iron Age hut.

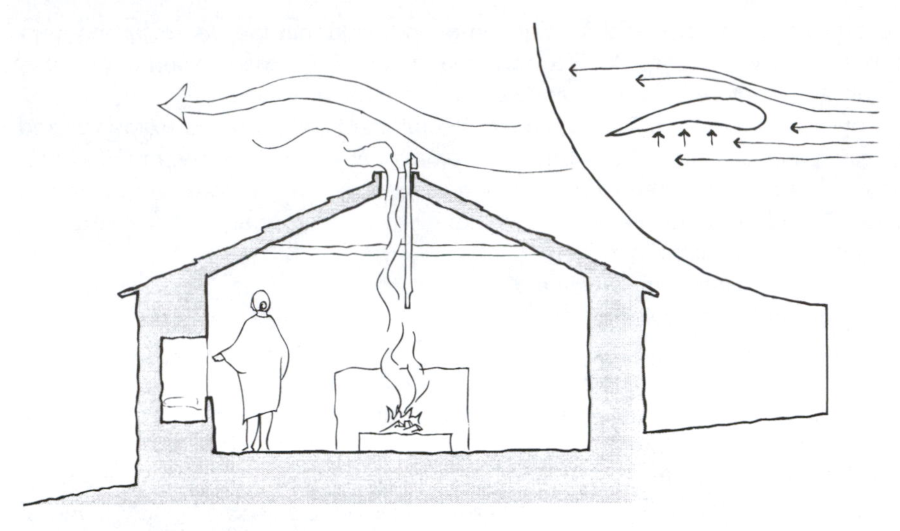

6.6 Kirbister Farm Museum.

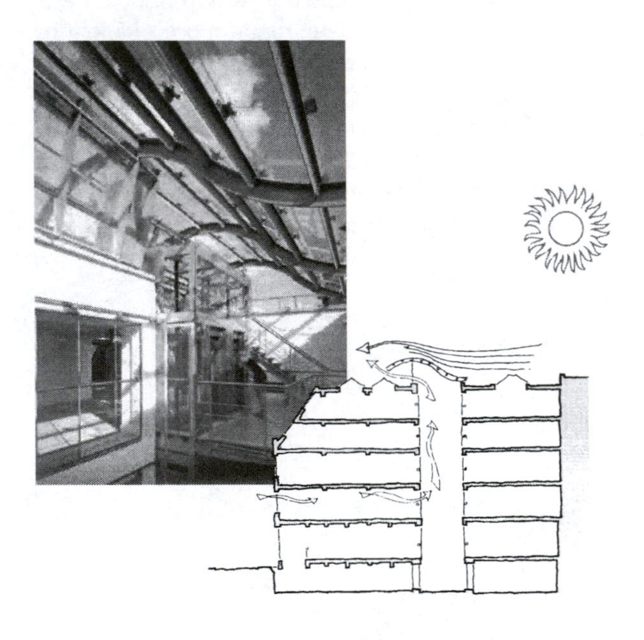

6.7 Offices in Stukeley Street, Covent Garden, London.

to increase the effect of the wind in extracting air from the atrium and adding to the stack effect. The architect likes to refer to the pioneer aviator Antoine de Saint-Exupery (perhaps best known as the author of *The Little Prince* – a little treatise on sustainability in itself) to describe how machines can engage humans poetically with the world. The juxtaposition of the old brick and concrete building and its new high-technology component suggests one particular sustainable approach.

Stepping back in time, it was in the eighteenth and nineteenth centuries that increasing prosperity and technological progress led to a burgeoning variety of building types and forms with their own special needs for ventilation. To take but one example, in Georgian Bath at the Ball Room in the Assembly Rooms by John Wood the Elder (see Fig. 6.8) where hundreds of people danced, for privacy the ground floor was windowless and lighting and extract ventilation were achieved by a clerestory. The main source of (relatively) fresh air was from the entrance doors off the main lobby. On a typical summer night it probably got extremely hot indicating that, although the art of society dancing was quite mature, the science of ventilating high-density, deep plan spaces was in its infancy. To afford themselves some comfort the ladies, mimicking honeybees, would fan themselves.

In the nineteenth century as people spent more and more time indoors and used more artificial lighting from combustible fuels, the need to remove the products of combustion began to be a major factor in ventilation. Fashionable New Yorkers in Manhattan townhouses, reliant on whale oil for lighting, had to get rid of the fumes just as much as the Inuits.

New engineering professions and industries were born to cope with the need for heating, ventilating and lighting these increasingly complicated buildings, and architects became more and more involved in the physiology, i.e. the services installations of their buildings. An excellent example is Barry's Italianate palace, the Reform Club in London (see Fig. 6.9), where in 1839 an advanced and well-conceived system using a fan driven by a five-horsepower steam engine was developed to ventilate the building and provide background heating.[7] Air was led through ducts carefully integrated into the building to be delivered into the principal rooms at ceiling level often at decorative cornices. It was extracted in a variety of ways including outlets above gas lighting devices and up the chimney flues.

6.8 The Ball Room, Bath.

6.9 The Reform Club.

The idea that buildings and cities needed ventilation i.e. that they needed to 'breathe', is not new. Many medical reformers in the nineteenth and twentieth centuries, including the father of the French writer Proust, were aware of the need for good ventilation to reduce the incidence of disease. Le Corbusier's interest in 'controlled respiration' for buildings (based at least in part on an obsessive interest in physical fitness) – was a product of a historical context and hardly sprang fully-formed like Aphrodite from the foam of the sea.

In some ways the natural successor of the Reform Club, both in terms of form and services, was the Larkin Building (see Fig. 6.10) of 1906 by Frank Lloyd Wright. This office building, sited next to the railway lines in Buffalo, New York was entirely sealed and mechanically ventilated to protect itself from pollution and noise – and so may be considered an important step in the development of the defensive building.

By this time mechanisation was healthy and well but it could be argued, at least in the case of Wright, under (his) aesthetic control. When Carrier (and others) developed air-conditioning for buildings all the elements were in place for the fully sealed mechanically ventilated, heated and cooled glass block discussed in Chapter 4 that was to come to symbolise an important strand of twentieth-century modernism. Whether beauty or beast, it consumed a huge amount of energy, produced vast amounts of carbon dioxide and was destined to perish. But what would take its place wasn't clear.

Previously, we have seen a number of examples of 'the next generation'. One of the most exuberant is the Queens Building at De Montfort University by Short Ford and Associates[8,9] (see Fig. 6.11) which developed, i.e. evolved from the earlier Farsons' Brewery in Malta, by the same architects.[10] Motivated by a desire to both announce the presence of a new university in an urban renewal area in Leicester and to minimise the use of energy, the designers turned to a highly articulated form to favour natural ventilation and daylight.

6.10 The Larkin Building.

The building, completed in 1992, is a study in how to use the internal organisation, external form and building fabric to reduce environmental impact. The architectural contrast with office blocks in a Miesian style (Fig. 4.7) could not be more marked. In place of a monolithic, heavily glazed block (albeit, at times an elegant one) there is a highly developed form with a well-insulated envelope which reduces the energy needs and also responds to a particular site on a particular (noisy) street in a particular city. High thermal mass is incorporated in extensive brickwork and concrete. The Queens Building is similar to a stationary, but very aware, animal (a bit like the resting camel of Fig. 5.1) that needs to maintain its temperature and organise its circulation systems and particularly its ventilation. In addition, it needs to capture significant amounts of light. (This is a common requirement for buildings but rare in the animal world where light is mainly used in specialised organs for vision – eyes. On the other hand, in the plant world there is of course a corresponding need for extensive light for photosynthesis – see Figure 7.24.) The Queens Building responds to the environmental challenge by first of all creating an urban city-scape on one side and a more articulated form for cross-ventilation and light on the other side which we see in Figure 6.11b.

To ventilate the deeper areas of the building such as the densely populated lecture theatres (see Fig. 6.11c) and eliminate any need for artificial cooling, air is brought in at low level at the perimeter and rises to exit from the dominant towers by the stack effect.

The terminations of the tall stacks create a new skyline in the area (in some ways similar to the towers of Italian Renaissance cities such as San Gimigniano) and show how forms in part derived from environmental considerations can be used to create architecture.

This is a 'natural' building and its detailing reflects this. In a mechanically ventilated structure one would have metal grilles to let air in and out and a powerful fan would be used to circulate air. Here, instead at the pedestrian level for the air inlets, clay tiles are used; thus a detail of the 'skin' suggests the whole and in particular a more biological approach to design – one based on understanding of functional (and aesthetic) requirements.

a

b

c

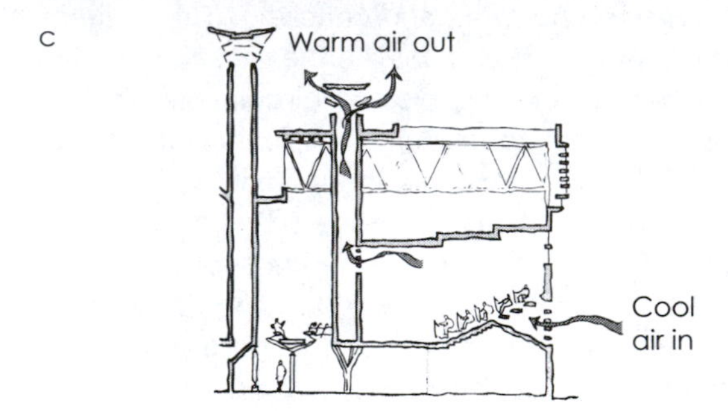

6.11 Queens Building, De Montfort University, Leicester.

a street view.

b 'back view'.

c section.

The ventilation is linked to the thermal mass of the building and at night lower temperature air is brought in to cool the spaces and prepare them for the next day. (The combination of ventilation and thermal mass is a sensible, efficient and economic way of regulating the internal conditions of buildings to provide comfort.) As the ventilation rate can vary with need, there are similarities with animal bodies. Human lungs, for example, 'pump' about 6l per minute at the normal resting level but can increase this to 25 times as much during strenuous activity.[11]

These considerations were taken further in the Contact Theatre by Short and Associates[12,13] which in responding to an exceptionally complex brief combines low energy and natural ventilation techniques with craft aesthetics.

Rather than imposing a preconceived idea of form (either from function, history or pure invention), the architect acts rather like the conductor of a chamber orchestra, bringing in the environmental engineer, structural engineer, other consultants and performers to make contributions to the design. The Contact Theatre is an extensive upgrading of an existing building, adding much-needed support facilities and more front-of-house space, whilst retaining the auditorium with its much-loved wide stage. The strategy adopted for natural ventilation also provided the opportunity for something of a 'makeover' for the theatre.

For the ventilation, one starts with a humble H-pot chimney ventilator as one finds, for example, helping to provide cosy, domestic winter comfort at Melsetter House (see Fig. 6.12).

Then one asks whether a similar form, magnified by a factor of about 1,000 (to allow for the larger volume as well as the much lower temperatures and thus energy densities of ventilation compared to combustion referred to previously) is suitable for a theatre for just under 400 people (see Fig. 6.13).

The answer is yes for a number of reasons. First, the Contact Theatre sits some way back from the main road of this part of Manchester so the enormous H-pot vents act as a distinct sign announcing the Theatre's presence. The overscaled vents appear rather like the tactic employed in Pop Art, magnifying a 'found' or 'ready-made' object to raise questions of perceptions. Second, the sizes of these forms were carefully calculated to ensure that the audience would be comfortable.

In some ways the Contact is another unusual immobile organism with certain features related to air and noise being the direct result of environmental pressures. The Theatre is located not far from a major noisy road and even closer to a noisy student

6.12 Melsetter House H-pot chimney ventilation.

a

b

c

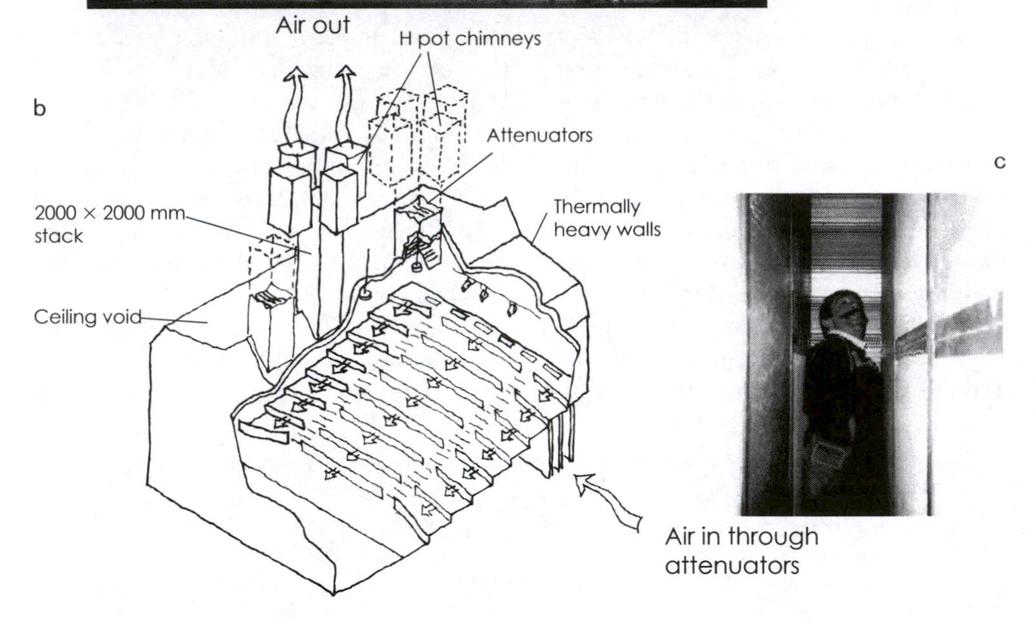

6.13 The Contact Theatre.
a front façade.
b cutaway axonometric.
c ventilation duct.

venue so it had to protect itself acoustically. This was done by setting the air inlets on the far side of the building and then taking the air through tall, deep and widely spaced acoustic attenuators (see Fig. 6.13c). This air entry is like a special elaboration of the surface – a 'stomata' with an acoustic role. (In the future we are likely to see further variations with openings that are finely tuned to levels of atmospheric pollution.) Air is then taken under the seats, through the main theatre and then out with assistance at times from low-energy fans. These were sized to deal with the peak condition of a full house at a brightly lit Christmas pantomime. In the exit stacks there are again acoustic attenuators to limit the noise getting into the theatre. Although control mechanisms are outside the scope of this book, it is worthwhile pointing out the similarity between natural biological systems and our (simple) mechanical ones. In the vertebrates, including, of course, man, the surface of the brain (medulla) is particularly sensitive to high CO_2 levels. The gas is monitored and when high there is increased ventilation of the lungs. At the Contact Theatre, the CO_2 level (as part of indoor air quality) is also sensed and rising levels cause the fan speeds to increase.

The theatre lobby has its own ventilation path. Here the fresh-air inlets are designed rather like sentry boxes made of bricks with clay air-bricks – these can be seen in Figure 6.13a in the lower left-hand corner. Avoiding the usual metal grilles for air inlets reinforces at the level of detail the perception that this building is not mechanically ventilated. A theatrical introduction to this theme is provided by the sheet-metal screen over the entrance that looks like a curtain lifted over a proscenium. The zinc that would normally be for the galvanised steel ventilation ducts in a theatre is here used as decoration and as a display of the workman's skill.

Heelis – the New Central Office for the National Trust

As a last example, let us look briefly at Heelis (see Figs 6.14 and 4.17), the new Central Office for the National Trust[14,15] by Feilden Clegg Bradley. The designers set out to improve on their work on the BRE Environmental Building (see Chapter 4) and create an even lower-energy office building. This was done in part by designing for natural ventilation in the summer (with mechanical ventilation with heat recovery in the winter), high levels of natural daylight and electricity from photovoltaic panels on the roof. The building also was well insulated and extremely tightly sealed in the winter. As energy conservation and the reduction of CO_2 emissions become more important the idea of being able to 'zip up' the building and provide only the ventilation required gains force. It is a bit like the Inuits who would be very wary of a 'leaky' igloo.

The starting point was to locate the two-storey building towards the south and so at an angle to the existing grid of surrounding buildings giving a long south facing façade and a suitable roof for PVs and north lights. An effect of 'layering', which adds to the richness of the architecture, is achieved at the south façade by the solar screen, a circulation space under the overhang and the main wall with its ventilation opening; the planting to the south complements this layering. To allow for natural ventilation and daylight in this deep plan, building wells (which are, in effect, small courtyards) were created. A daylight factor (see Glossary) of 15 per cent was achieved on the first floor and 5 per cent on the ground floor helping to create a feeling of a light, airy space. Glazing, of course, transmits both light and heat and so there is a fine balance to be achieved to ensure thermal comfort in all seasons. In the summer fresh air is brought in from the perimeter and from the wells and exits mainly via high-level stacks which can be seen in the figure. Night-time ventilation with cooler air in conjunction with an 80 mm concrete slab to provide thermal mass keeps the building comfortable. The roof at Heelis works quite hard and does this in perfect co-ordination with the section – it is a source of visual interest, lets in light, allows ventilation (and heat recovery), produces electricity and stores heat and coolth in its thermal mass. The importance of this mass has its biological analogy in, for example, the camel, where it is used as a thermal modulator preventing rapid rises in body temperature when exposed to high

a

b

6.14 Heelis, the National Trust headquarters.
a View of the south façade.
b View of the north-east façade.

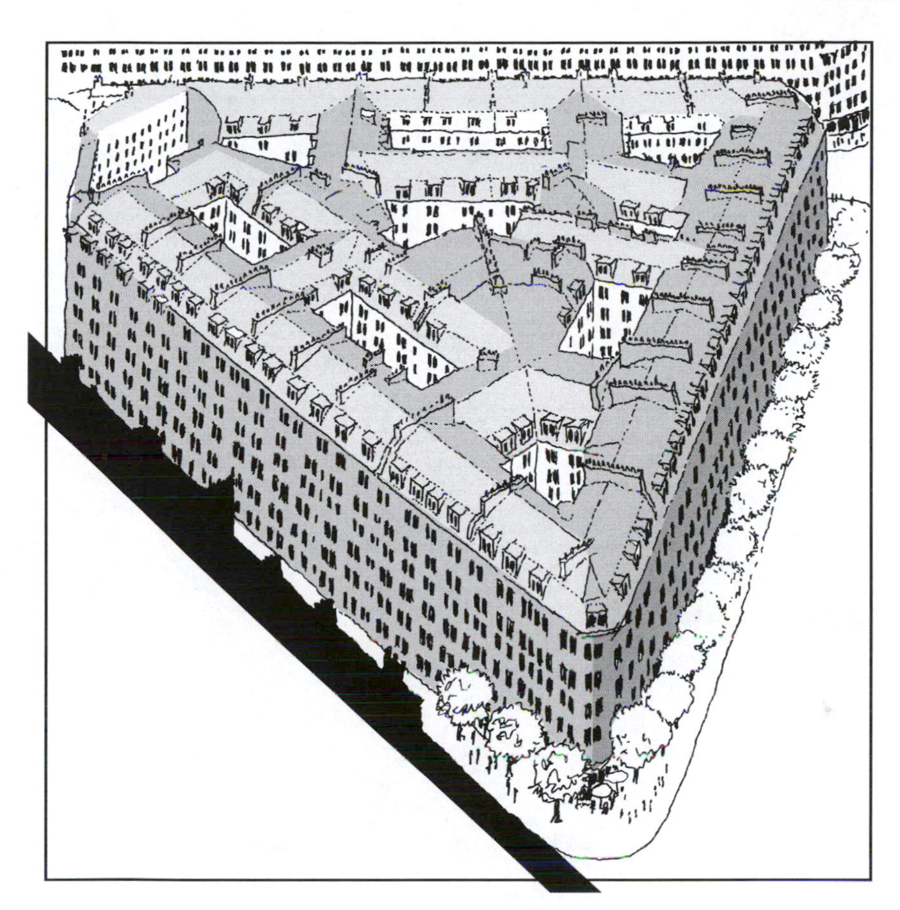

6.15 Nineteenth-century housing with courtyards.

solar radiation. (In the polar bear the high thermal mass helps provide protection in winter against a fall in temperature.) In the winter a mechanical ventilation system recovers heat from the extracted air and transfers it to the incoming fresh air. All in all, this building operates like a complete system, similar to an organism, with surface, structure, materials and services (i.e. physiology) complementing each other.

The importance of morphology is a recurring theme in this book and it is interesting to compare Heelis with its wells with a particularly striking block in an area of Paris planned by Haussmann in the latter half of the nineteenth century (Fig. 6.15). The courtyards are in fact deep wells serving similar purposes of providing ventilation and daylight. And, of course, they make us think of the Matmata dwellings of Chapter 2.

Heelis with its deep plan structure punctuated by courtyards also points us towards a sustainable urban block, as can be seen in the sections of Figure 6.16 which compare various forms. The block could be designed for solar energy and could incorporate both thermal and rainwater storage for recycling at basement level.

Heelis shows that form, space, light, ventilation, energy and architecture can be conceived of as a whole. This can be developed into an approach to deep-plan buildings and urban design which both reduces the demands on the global environment and contributes to the local sense of place.

Heelis National Trust central office

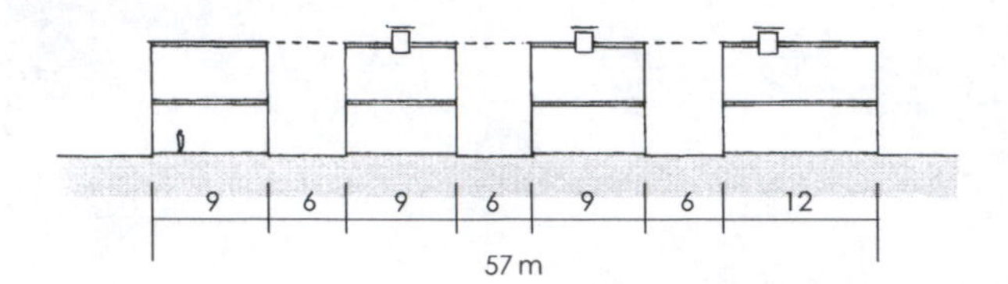

London courtyard housing

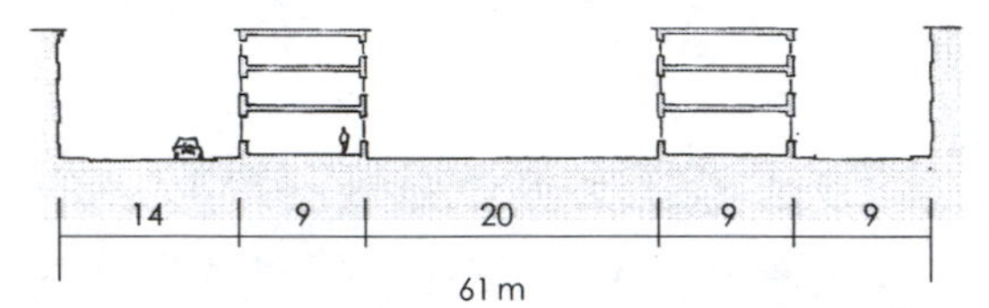

Sustainable urban block

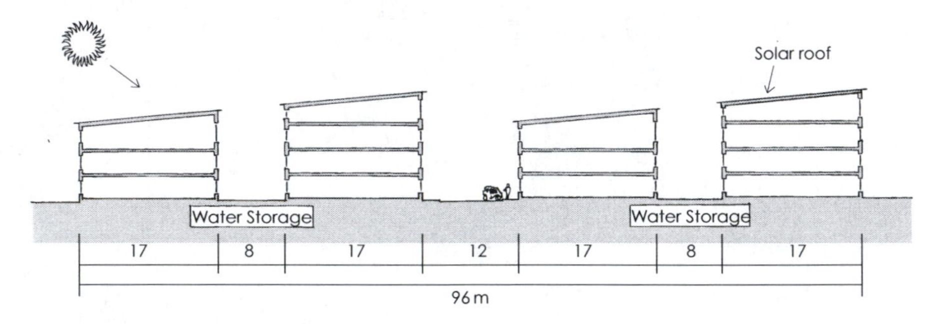

6.16 Comparison of various forms.

Further reading

Von Frisch, K. (1974) *Animal Architecture*, London: Hutchinson & Co.

Building references

Offices in Stukeley Street: Architects – Jestico and Whiles; Environmental Engineers – HGS; Structural Engineers – Price & Myers.

The Queens Building: Environmental Engineers – Max Fordham LLP; Structural Engineers – YRM/Antony Hunt Associates.

Contact Theatre: Environmental Engineers – Max Fordham LLP; Structural Engineers – YRM/Antony Hunt Associates.

Heelis: Environmental Engineers – Max Fordham LLP; Structural Engineers – Adams Kara Taylor.

7

Light and shade

Introduction

Light opposed to darkness and land rising from primordial waters – these are the fundamental components of many creation myths from all over the ancient world.

> In the beginning God created the heaven and the earth. And the earth was without form, and void; and darkness was upon the face of the deep.[1]

On the first day, so the Biblical creation story goes, He created 'material light', which took the particular forms of the sun, moon and stars, and at the same time He divided light from darkness and day from night. In her book *Man and the Sun*, Jacquetta Hawkes ponders the mystery of human consciousness eventually evolving from inorganic matter, and how the significance of sunlight for terrestrial life must have figured large in the development of myth, religion and art. We are, she says, the 'children of light and the children of darkness. This imagery dwells in every one of us, constantly coming to our lips, and flooding out in poetry and music'.[2]

The biological explanation of life's mysterious emergence has some correspondence with the mythological accounts. One theory is that the 'spark' of life might have been produced by the reactive potential of ultraviolet light on gases such as hydrogen cyanide, methane and ammonia that were dissolved in the primordial waters.[3] The ultraviolet light would have had much greater effect on the Earth then, when life first evolved because the atmosphere contained virtually no free oxygen, hence there would have been no ozone layer. The development of most life forms only became possible as oxygen levels increased and a protective shield against ultraviolet developed (see Appendix A). Sunlight was the agent for life.

As we saw earlier, Le Corbusier came to see the reciprocity between light and dark as an expression of the fundamental principle of life, something that came to shape his design thinking after the Second World War. Two drawings from his *Le Poème de l'Angle Droit* (1955 – translated as *The Poem of the Right Angle*) vividly convey how this poetic notion connected his architecture with the fundamental relationship that exists between the sun and life on Earth. Beneath a simplified section of the Unité d'Habitation with the sun's path inscribed above, is depicted a seed in the Earth, the seed sprouting, the root going down into the darkness, the plant stem rising to the light, as plant life always will (Fig. 7.1). Another drawing shows the diurnal rhythm of night and day in relation to the cycle of the year, depicting the relative darkness of winter and the long light days of summer (Fig. 7.2).

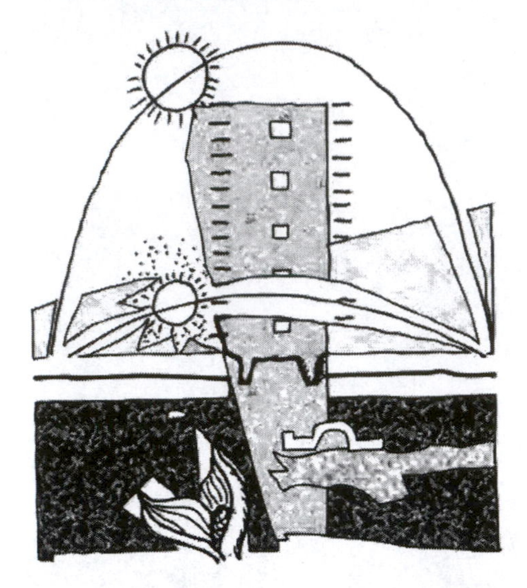

7.1 Light and dark, sun and seed. (After Le Corbusier's lithograph in Le Poème de l'Angle Droit.)

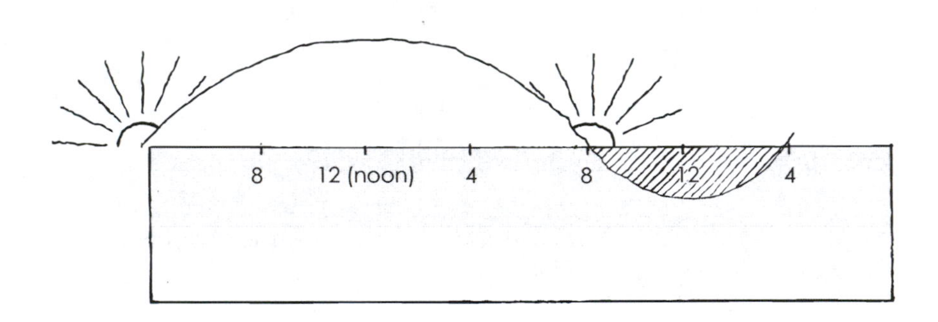

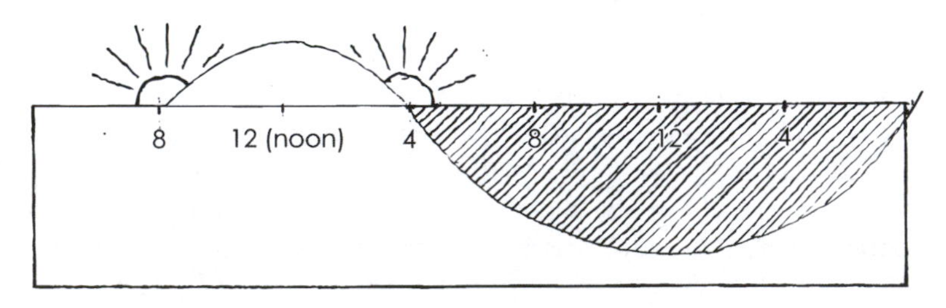

7.2 Diagram graphically indicating diurnal and seasonal rhythm of day and night, light and dark. (After Le Corbusier's drawing in Le Poème de l'Angle Droit.)

The acclaimed writer on comparative mythology, Joseph Campbell, has suggested that the diurnal rhythms of night and day, darkness and light, and seasonal equivalents of winter and summer, are the constant background reminders of our deep human connection to the Earth. Life responds to, or is orchestrated by these rhythms. The built environment should aim to enhance human interaction with the natural cycles of light and dark as far as possible. As well as this poetic consideration, there are important practical reasons for designing with daylight in mind. These can be summarised as, first, aiming to optimise the distribution of daylight into the interior of buildings, consequently reducing the amount of energy used in artificially lighting interior spaces, and second, to ensure there is the appropriate level of light for visual comfort. As we saw in Chapter 4, the development of the deep plan in commercial buildings has led to an extensive dependence upon artificial light to produce a satisfactory working environment throughout the day. Careful thought and imagination is required in designing how best to invite sunlight and daylight into buildings.

We earlier discussed a basic morphology of built form and this can be re-visited in relation to light (Fig. 7.3). The hole-in-the-wall window was the first purposely designed source of light, buildings before being lit only by a smoke-hole or light from an open door. This arrangement allowed the single-cell house to evolve, but effectively determined that the basic form of building would be two rooms deep, each room requiring an external wall for a window. With the development of glass, the rooflight introduced a new possibility. Because the sky is the source of illumination, the inclined surface of skylights allows more light to penetrate than a window. A large covering of glass over what was effectively an exterior courtyard – as at the Oxford Museum – introduced the glazed atrium and allowed buildings to become more than two rooms deep – as we have seen, for example, in the Heelis Central Office for the National Trust. A variation of this is a saw-tooth roof which can accommodate north light whilst excluding direct sunlight. This basic morphology of built form could be likened to the basic grammar of language, poetry being the aspiration to weave more evocative forms of expression, whether in words or with light in the case of building.

Over the past few years there have been many technical innovations to enhance the penetration of light into buildings. But recent research came to the conclusion that a 'simple system (roof and façade for instance) was found to perform better than advanced façade systems that attempted to divert diffuse daylight deep into the building'.[4] This chapter outlines some traditional and inventive ways by which light was invited into buildings and reviews how new technologies might be most effectively and appropriately incorporated.

The nature of light

But what is light? Purely physically, the answer is straightforward. Light is the visible portion of the electromagnetic spectrum that covers everything from X-rays to microwaves. The energy from the sun is in the 'middle' of that spectrum, and light is roughly in the centre of that with ultraviolet (UV) to the left and infrared to the right (see Appendix A). However, there are also subjective aspects of light. For example, the adjustable sensitivity of the retina means that the appropriate brightness of an object depends upon its surroundings, which is critical to architecture and the arts.[5] The physical, objective side and the subjective are often intertwined.

Differences in spectral composition and intensity are important in creating a

a. Light from smoke hole and window
b. Lantern light
c. Rooflight
d. Glazed roof
e. Large span structure with rooflight
f. Saw-tooth roof with north lights

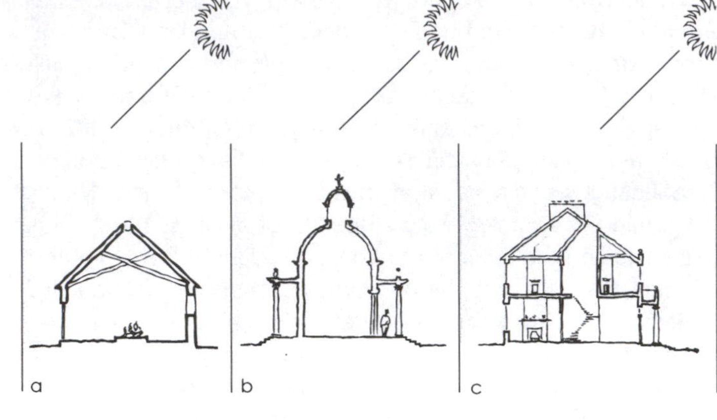

specific sense of place, as are the angle of the sun and the nature of the reflecting surfaces that the light strikes. These include both the Earth's surfaces and reflections from the underside of clouds. The tranquil beauty of Dutch landscape paintings, for example, is due in part to the painter's ability to capture the specific light of the lowlands. We are sensitive to many different types of light, both inside and outside.

Although our mental image of light is associated with a visible sun, all too often clouds obscure the sun. Our architecture needs to succeed visually in these overcast conditions as well as in sunlight. This involves sensitivity to materials, colour and form. In low light conditions there is little contrast, so the fine grain of materials and colour harmonies become more important. In the UK, under overcast skies one sees the advantages of the colour white and cream-coloured limestones, of green vegetation and light grey slates; the best solutions can often be seen in garden design. In more sunny climates, architectural forms from cornices to deep reveals to colonnades, in fact anything that makes the surfaces more complex or the forms more interesting when the sun shines on them, can be more appreciated. Traditional Japanese architecture was acutely aware of this. Japan is often overcast and sometimes sunny. In his classic work *In Praise of Shadows* (1933), Tanizaki talks often of a degree of dimness, of finely grained wood, of a pensive lustre and of how lacquerware can only be appreciated in dim half-light. 'The beauty of a Japanese room depends on a variation of shadows – it has nothing else.' This in part is created by the architectural forms, for example, by extended eaves or a verandah. Tanizaki goes so far as to suggest that if the Japanese had developed their own physics and chemistry (instead of adopting Western science in the latter half of the nineteenth century) then the facts they were taught – such as the nature and function of light – might have presented themselves under other forms. And we agree with the sentiment of this; there are many ways of seeing.

Light and shade should be considered together and this is both true for buildings and the city. The play of sunlight entering a room from several sources often thrills us, and the light filtering through plane trees in London parks is one of the delights of the capital.

Glass, of course, transmits heat as well as light and so glazing is a complex and subtle question of balance. Too little glazing and the need for artificial lighting increases in order to achieve the light level that is needed (see Appendix A). Too

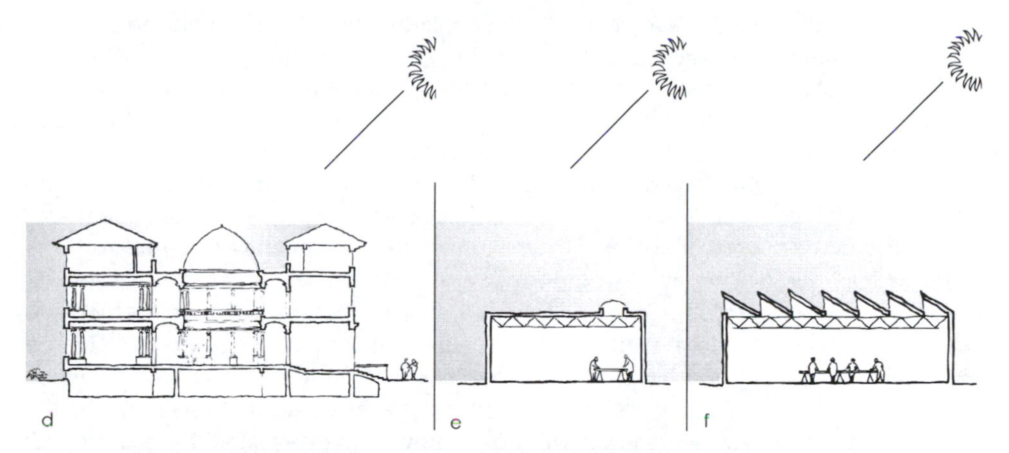

7.3 Diagrammatic sections showing morphologies of built form in relation to obtaining daylight.

much glazing leads to high heat losses in winter, although this is slightly less of an issue as we move towards well-insulated, multiple-glazed solutions. In the summer, too much glazing leads to high solar gains, overheating and the need for cooling as well as potentially causing glare.

Moonlight is about 0.2 lux, an overcast sky about 5,000 lux, and our brightest, sunniest sky about 100,000 lux, which corresponds to, say, 1,000 W/m^2. If 70 per cent of this, about the heat from a rather crowded seven people per m^2, is transmitted into a room it can be fine in the depths of winter but excruciatingly hot in summer – as was noticed at Le Corbusier's Salvation Army Hostel. Comfort in building requires a holistic approach that considers the control of solar gains, the use of the fabric of the building, and in particular its thermal mass and ventilation together.

Interest in buildings requires areas of opportunity, of tension, and of excitement. Perimeters are boundary conditions that can function in this way. As we saw in Chapter 6, perimeters can have specialised areas to let air in and out. In cities, the edge onto the street may have a particularly fine grain, as for example in London's Covent Garden, with its mix of restaurants and shops. In buildings, the perimeter with its high light levels and views of the outside world is attractive (as long as solar gain is controlled). In Germany, the health and psychological benefits of natural light are recognised by legislation that requires the design of office working spaces to have direct communication with the outside.

In the next few sections we will look at a number of historic and modern examples of approaches to light. The earlier examples are, of course, from a time when daylight was all-important. Artificial lighting until recently has been rare but its impact has at times been exceptional.

The technological invention of oil-burning lamps in the Magdalenian period helped, amongst other things, painters to venture deep into the caves at Lascaux to astonish us thousands of years later with their bison, mammoths and horses.[6] In the Middle Ages the glow from the fire and candles extended the length of the sociable (and working) day. In Georgian Bath, candles along with gold-rimmed plates and glasses were used as a display of wealth. Coal gas (approximately 70 per cent hydrogen, 20–30 per cent methane, 7–17 per cent carbon monoxide and a few other gases) was used for both external and internal lighting in the UK from the early part of the nineteenth century until the early part of the twentieth when it was displaced by electric lighting (even though many people preferred gas on aesthetic grounds).[7] Whale oil (a non-sustainable source if ever there was one) was commonly used for lighting on the Eastern seaboard of the USA – in large part due to whaling ships being based at Nantucket – as was coal gas in the nineteenth century. However, the discovery of oil in Pennsylvania in 1859 (noted previously) led to a growing supply of 'kerosene, the cheapest illuminate ever known'.[8] In the United States also, of course, electricity became the prime source of artificial illumination when first made available there.

The widespread availability of electric light in the industrialised world in the twentieth century transformed domestic and working spaces, but, as we have learned too late, at a significant cost to the environment. A sustainable architecture (and the Heelis National Trust building is a good example) will need to find a new balance between natural and artificial light that reflects this inside the spaces we make and on the façades. In the next few sections we will look at how some designers have dealt with lighting in their times and how this has influenced architectural form.

Domestic places

Good daylighting design requires a careful consideration of both the qualitative and the quantitative factors associated with light. The optimum amount of light for any particular activity can be measured in design by calculation and models, but the mood created by light – which is possibly the primary intuition of the meaning of a space – is just as important. The representation of light in space and the way it falls upon things has been a major preoccupation of painters, reflecting the profound significance light has for the human eye and the interior of buildings.

Light was particularly important to the seventeenth-century Dutch School of painters, particularly those masters of domestic interiors such as Vermeer, de Witte and de Hooch. Not only do the paintings of de Hooch depict the poetic qualities of light, they also show how the quantity can be adjusted. In his 'An Interior, with a Woman drinking with Two Men, and a Maidservant (*c.*1658)', we see the men sitting at a table beside a window to which a young woman is holding up a glass of wine for their appreciation (Fig. 7.4). Setting the table by the window seems apt for this domestic scene. The glass of wine held up to the light is immediately grasped by the viewer as an act of clarification and intensification. The gleam of glass and the transparency, glow and colour of the wine are all enhanced by taking the glass to the light.

The shutters at the window show how the quantity of light could easily be adjusted in this room. Dutch townhouses of this period occupied deep and narrow plots, hence the need to maximise light penetration. This was achieved by having tall and wide windows on the street façade – as depicted here by de Hooch – and often with small courtyards further back (as can be seen in his 'A Boy Handing a Woman a Basket in a Doorway', for example). The shutters shown in the painting in Figure 7.4 are hung in such a way as to give a wide range of light levels and degrees of privacy. All the shutters could be opened to give maximum light; the bottom shutters could be closed to give privacy but plenty of light would be distributed across the room from the top windows; or the bottom shutters could be open and the top closed to intensify the contrast of light and darkness in the room, for the particular pleasure of a painter perhaps. The shutters would have also been important for controlling the amount of solar gain or heat loss. (See Appendix A for a more detailed discussion.)

7.4 Pieter de Hooch, 'An Interior, with a Woman drinking with Two Men and a Maidservant'.

A general point this illustrates is that the amount of light that enters a space depends on the area of glazing, its type and its disposition. It is probably best to aim for the maximum amount of daylight, subject to considering glare and excessive solar gain. In quantifiable terms, lighting for work tasks should be 500 lux (see Appendix A).

Frank Lloyd Wright said that 'Proper orientation of the house ... is the first condition of lighting [a] house.' If the site is not an urban one and allows full scope for this then, as we saw in Chapter 3, the rooms of a building can be distributed according to the most appropriate orientation. Just as Wright's use of central heating increased his scope for making more free-flowing spaces, so it prompted a finer consideration of light. 'Daylighting can be beautifully managed by the architect if he has a feeling for the course of the sun as it goes from east to west and at the inevitable angle to the south.'[9] His Prairie Houses are characterised by low overhanging eaves, which reduce the amount of direct sunlight and skylight (as well as solar gain). This is compensated by the wide windows that fitted Wright's modern conception of space with its potential for the exploration of different qualities of light. A general principle that can be deduced from his houses is that planning a building with highly articulated spaces and an extensive perimeter affords plenty of opportunity for introducing daylight.

We saw this in the more traditional Melsetter House where Lethaby located the dining room and the bedrooms on the east to catch the morning light. The extensive perimeter is particularly noticeable in the drawing room, which was given a projection to the south such as to be in the sun's path for most of the day. The shape, size, type and position of the windows have a significant effect on the quality and quantity of light notwithstanding the orientation. For example, the drawing room at Melsetter House has three tall windows on the east, two on the south, and a small, high-level window on the west wall (Fig. 3.15). This gives an even, balanced quality to the light as well as offering points of sunlight throughout the day (Fig. 7.5).

In contrast to this, Charles Rennie Mackintosh has a single, south-facing bay window to light the drawing room at Hill House (1902–03, Fig. 7.6). This provides an intense box of light against a relatively dark room. Holes cut into the supporting posts show bright against the post's shadowed side and reinforce the idea of capturing light. The bay window is almost an adjunct to the room itself allowing the occupants to

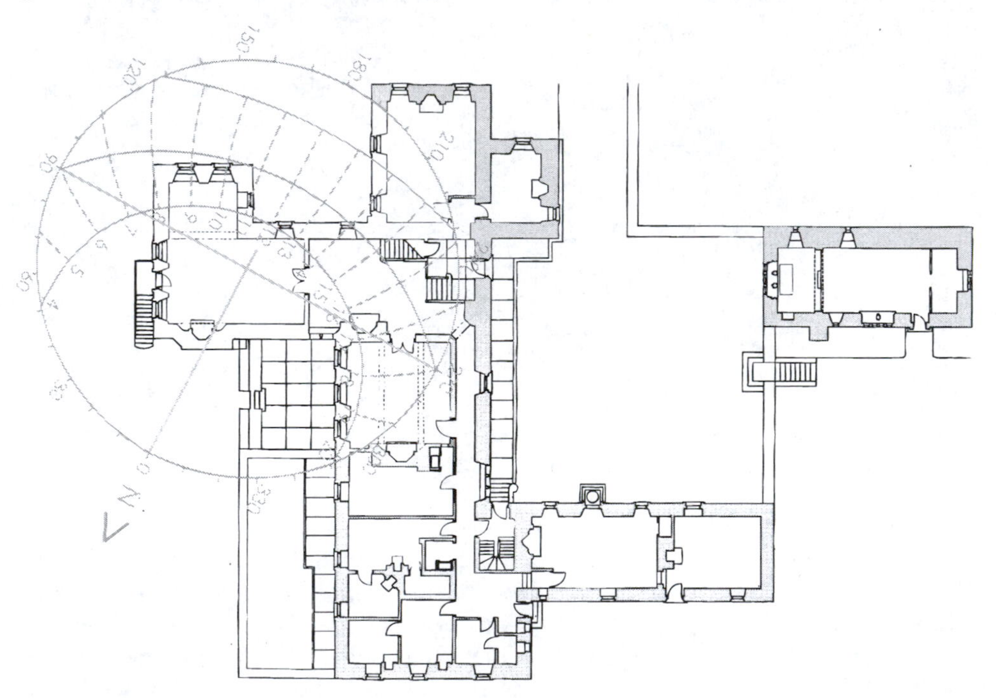

7.5 Ground-floor plan of Melsetter House with sun path diagram superimposed over drawing room. (Note that the sun path diagram is for latitude 51°N, whereas Melsetter is 58° 47′N, consequently the sun rises and sets further north in summer – giving longer days – and further south in winter – giving shorter days.)

7.6 Drawing room at Hill House, Helensburgh, by Charles Rennie Mackintosh.

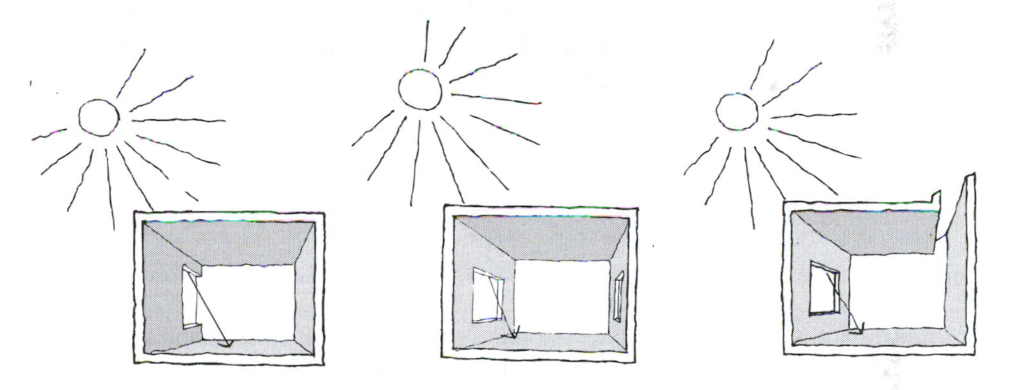

7.7 Diagram of how sunlight can be reflected off walls to illuminate rooms. (After Charles Moore's drawings in The Place of Houses. Some of Moore's arrows have been omitted here.)

move from a relatively low-lit space into a space almost as bright as the day itself, a veritable box of light. Although this arrangement gives a very poetic or dramatic effect of light, it does illustrate the problem of glare, which a large area of glazing can produce.

In his book *The Place of Houses*, the architect Charles Moore showed how the section of a building could be manipulated to reflect light into rooms. An important consideration for Moore was to avoid great contrasts of light and dark in a room (such as at Hill House) that produce glare. By positioning a window in one corner, for example, light might be reflected into one particular area of the room. His more elaborate sections show how light can be introduced in a number of ways to balance the levels of illumination – his arrows indicating the nature of sunlight are more suggestive than objective (Fig. 7.7).

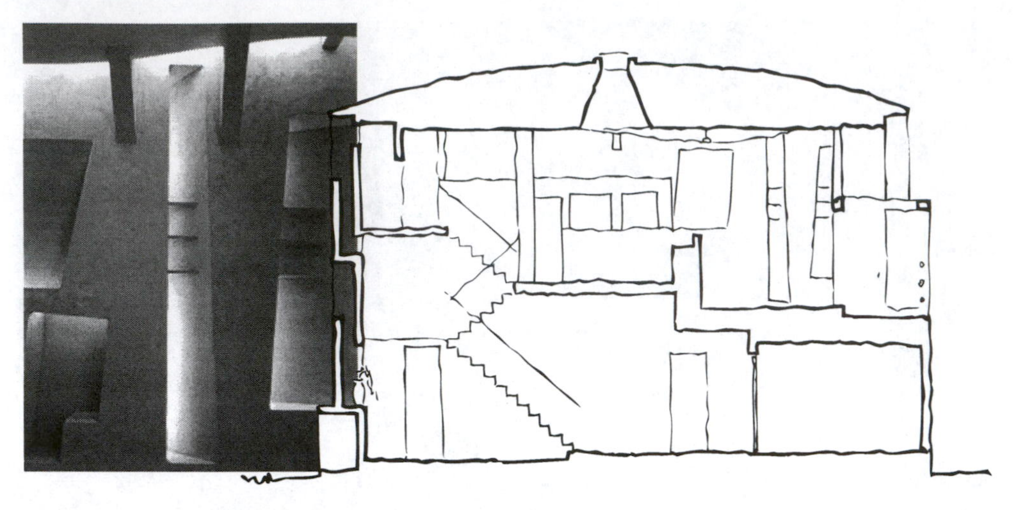

7.8 Section and detail of the Chiaroscuro House, Tokyo, by Ushida Findlay.

A particularly striking illustration of this approach can be seen in the aptly named 'Chiaroscuro House' (1992) by Ushida Findlay. Inventive use is made of a thick wall constructed from a double-skin timber frame, the inner and outer members separated by 40 cm. The space between the inner and outer skin is 'hollowed out' in places to channel the light to where it is required. This operates in both plan and section to produce very dramatic light, boxes of light within the solid wall. The source is often unseen, and light is funnelled down not unlike as in a periscope (Fig. 7.8).

Skylight

Georgian townhouses of the eighteenth century developed deep plans on narrow frontage sites, which prompted the introduction of skylights above the staircase. A relatively small area of glass allowed a good deal of evenly spread daylight down three or four storeys of the stairwell, as can be seen in the section of Sir John Soane's house at 13 Lincoln's Inn Fields[10] (1810–23, Fig. 7.9). The stair itself was wrapped around the perimeter to create a void at the centre down which the light could fall. From the latter part of the eighteenth century and particularly the first half of the nineteenth century, extensive top-lighting transformed the interior of buildings. This was made possible by a more sophisticated scientific understanding of the chemical processes involved in glass making leading to more efficient and cheaper production of glass. (The Pantheon, which Soane had measured and drawn, had been lit only by a rooflight – a circular open hole without glass at the apex of its dome – for nearly 2,000 years (Fig. 7.10)).

Glass had been in use for a long time. Glass beads found in Egypt date from around 2500 BC. But the first use of flat glass probably dates from the Roman era, for bath-house windows were discovered at Pompeii. Glass blowing (which probably originated in Syria in the first century BC) allowed the production of flat sheets by blowing the bubble into a mould. Like many technical processes, glass making lost its way after the collapse of the Roman Empire and clear, colourless glass was difficult to produce in the early Middle Ages. In the later Middle Ages, the process of making

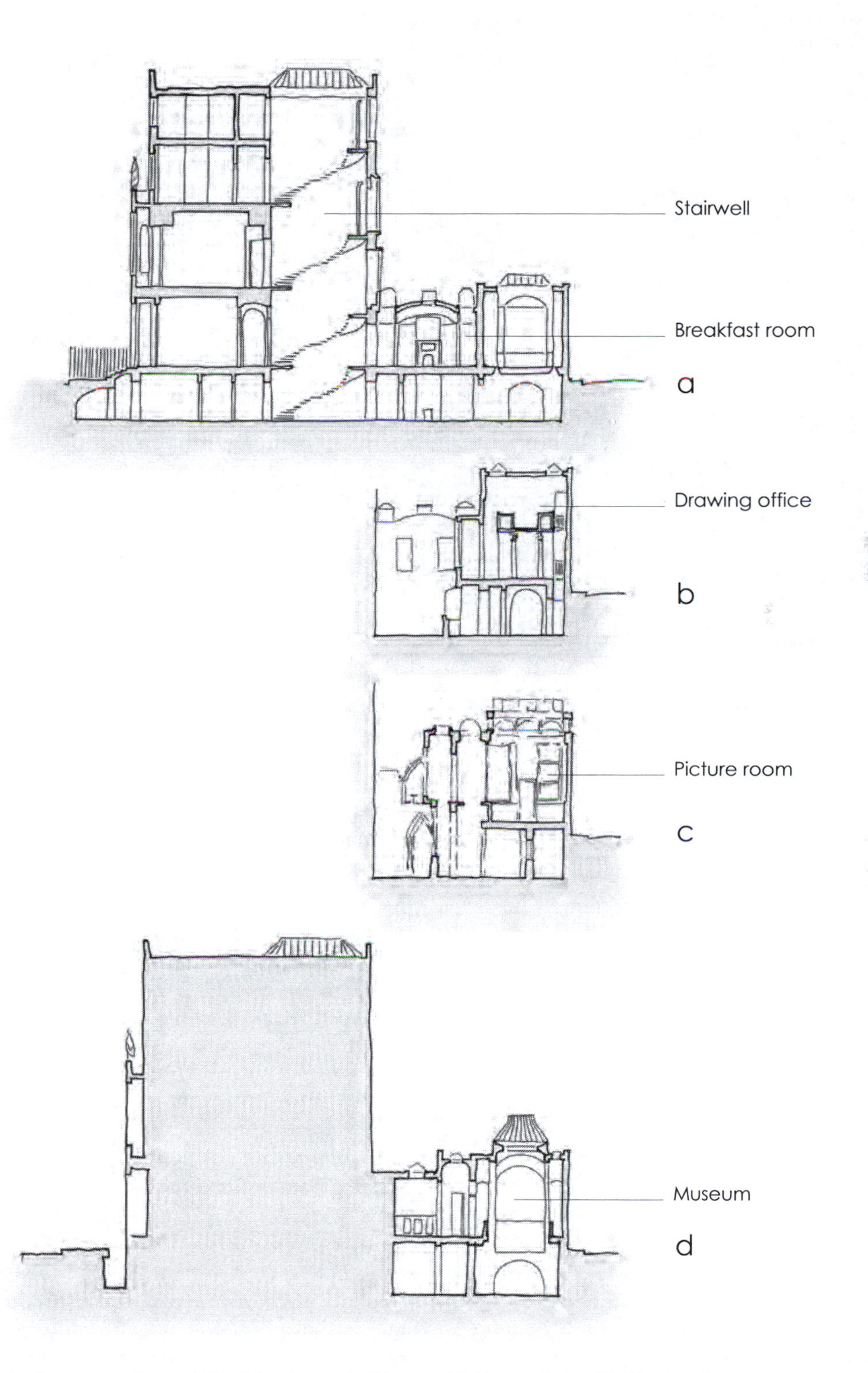

7.9 Cross-sections of Sir John Soane's House, 13 Lincoln's Inn Fields, London.

7.10 The Pantheon, Rome.

'crown glass' was developed and this remained essentially the same until the nineteenth century. Crown glass was produced by cracking off the blown bubble, forming a conical shape, inserting a rod, returning it to the furnace and spinning it until a flat disc appeared. Spinning generated a centrifugal force that spread the glass until a circular sheet formed. This could then be cut into rectangular pieces as required. (The 'bull's-eyes', a popular choice for cottage windows today, was then the least desirable piece of glass.) In 1832 Lucas Chance began to produce plate glass, making possible larger panes, and flat glass production machinery was introduced at the end of the nineteenth century.

Soane exploited the development of rooflights to the full. An array of skylights, often the opening lined with mirrors, allows light to penetrate into his four-room-deep town house, which had no rear aspect. Several rooms were shaped for specific purposes when he transformed the house into a museum. Twenty-four rooflights at the rear of the house ensure that the level of illumination is high. Some rooms have no window so a rooflight was essential, but each room is given a particular character, by a particular quality of light.

The breakfast room has a lantern light over the centre of its domed ceiling as well as a rooflight lighting two opposing walls, which light paintings below. The source of the light is not directly seen from most positions in the room, which adds intrigue to the illumination (Fig. 7.9a). There is also a window to a small light well, the combination of four light sources making a very bright and evenly lit small space.

Immediately beyond this tiny room is a triple-height, lantern-lit space known as the museum where Soane's collection of architectural fragments is housed (Fig. 7.9d). Soane designed a tall, cone-shaped lantern such that direct sunlight would be admitted between noon and 6 p.m. in the summer months, sunlight being desirable to set off the white marble architectural fragments and sculpture in this space.[11] Near the top is a tiny internal window opening to his drawing office and at the bottom are a sarcophagus and burial urns; the light of heaven (and creativity) contrasted with the shadow of death.

Located in the same rear zone of the house, the picture room is lit by clerestory lights on three sides, to which Soane added a central linear rooflight (Figure 7.9c). This was in white, acid-etched glass to balance the light in the room better.[12] Soane shared a deep interest in light with his friend the painter J.M.W. Turner. In his Royal Academy Lectures, Soane advised students to 'reflect upon the different modes adopted by Painters of introducing light into their studios'.[13] Turner had built his own gallery that incorporated rooflights draped with muslin to filter light through. Soane emphasised drawing as a technique to understand light, but also used models with opening flaps to see and test the effects of light inside spaces he designed.[14] By the end of the eighteenth century, when Soane began to practice, more sophisticated, scientific understanding of light was becoming available than the rudimentary, empirical rules set down in Renaissance and later architectural treatises.[15] He made use of this knowledge as well as the range of design techniques just mentioned in order to create what he called 'the poetry of architecture' through effects of light.

The Soane Museum is an object lesson in how light can be introduced to rooms where there is little scope for windows. In his studio/drawing office the section is manipulated to bring light down the walls on either side of the space to fall on the head of a drawing board (Fig. 7.9b). This provided optimum conditions for the draughtsmen's work in an age before the electric angle-poise lamp. The studio/drawing office is rather like a large table itself set into a double height space. Pulled slightly away from the walls, this arrangement allows light to penetrate down to the floor below. Before electric lights, this kind of consideration was essential to design. Even in the era of gas light at the end of the nineteenth century, Charles Harrison Townsend used a similar section at the Whitechapel Gallery (1901) to bring daylight down to the ground floor of a two-storey building where no side or rear windows were possible (Fig. 7.11).

One of the problems when skylights are used for galleries is preventing sunlight striking displays sensitive to it. In the Dulwich Gallery (one of the first public art galleries designed in 1811) Soane incorporated a simple yet clever device to control this. The gallery is lit from central lanterns originally covered with lead but having glass clerestory windows on each side (the opaque roof has subsequently been replaced with patent glazing). A small detail here allowed Soane to adjust the amount of light when the building was almost completed. Simply by extending or shortening the cill where the rooflight opening penetrates the ceiling would reduce or increase the amount of light in the gallery below (Fig. 7.12). Extending the cill would be particularly useful and easily done if it was found that sunlight would fall on the paintings.

A more contemporary imaginative, yet simple, solution to this problem can be seen at van Heyningen and Haward's Sutton Hoo Treasure Ship Museum (2001). Central lantern lights with a canvas awning suspended beneath (Fig. 7.13) light a dark and cavernous space. This not only prevents sunlight penetrating down onto the displays

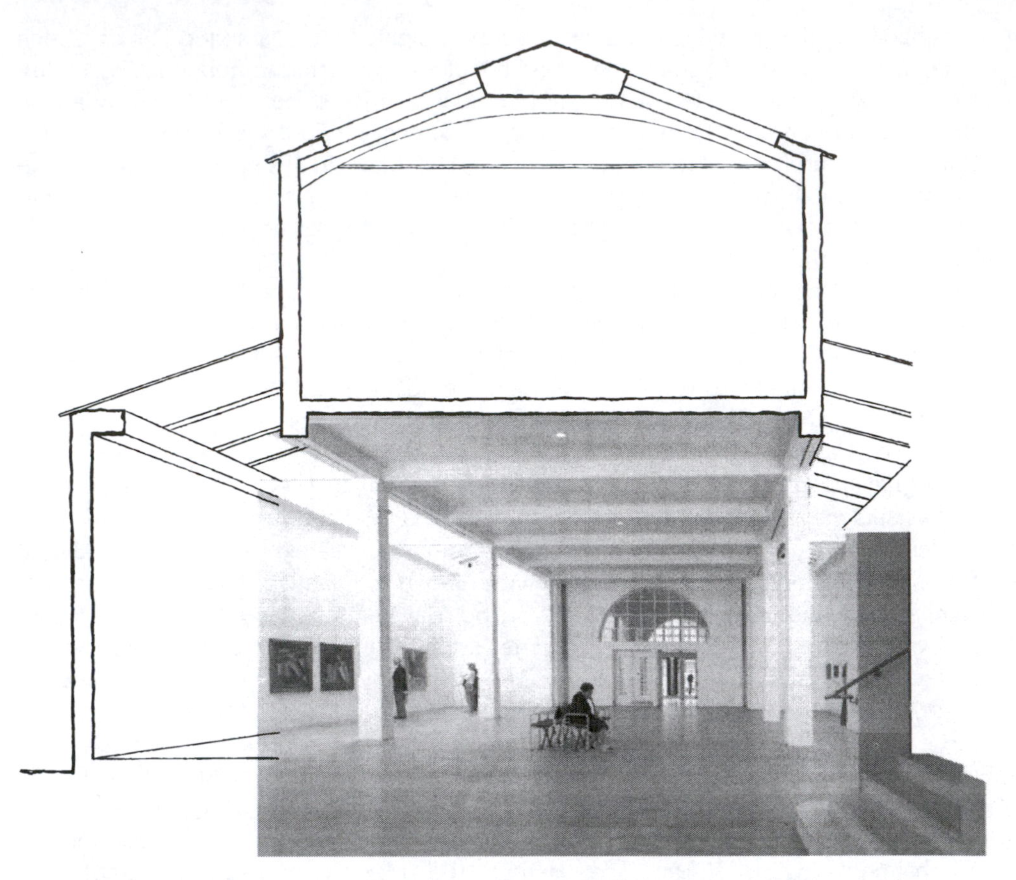

7.11 Cross-section through the Whitechapel Gallery, by Charles Harrison Townsend.

but also has the added benefit of making the light itself visible, sunlight glowing in the white canvas in the roof above. This approach has the potential for regulating the amount of light entering a space in a low-tech way, rather as a ship's sail can be trimmed to take more or less wind.

It was the development of the glasshouse, or greenhouse, in the eighteenth and nineteenth centuries that paved the way for glass-roofed buildings. Greenhouses or hot houses had been known in classical antiquity, but these were probably more like the eighteenth-century orangery, with many and tall south-facing windows, than the glasshouse or conservatory we are familiar with today.[16] These probably originated in the 'forcing frames' used by gardeners in the eighteenth century. Scientific studies showed that inclining the glass approximately perpendicular to the angle of the sun's rays increased the amount of heat gained.[17] The same occurs today and we tilt our photovoltaic panels at an angle based on the solar irradiation throughout the year to increase the energy gained.

The development of iron structures, industrialised production and cheaper glass led to the glasshouse becoming a major building type in the nineteenth century, and the technology was transferred from increasing heat to gaining light in railway stations and most famously for the Great Exhibition of 1851. (The building became known as the Crystal Palace.) These developments led to the design of buildings with a glazed

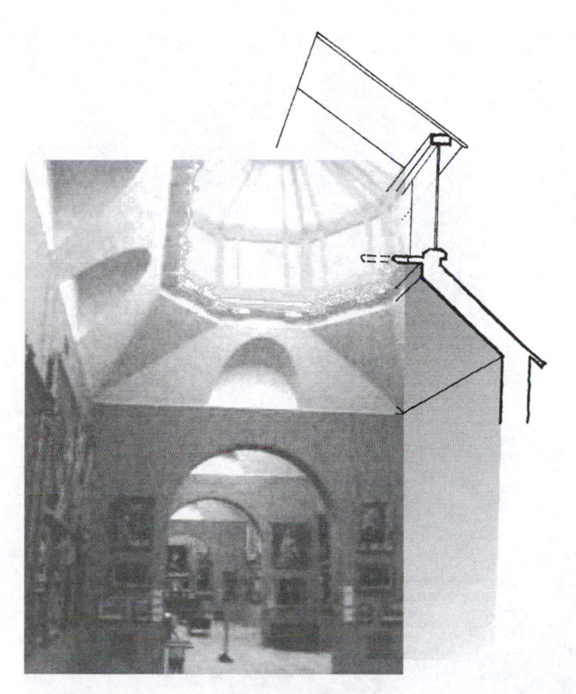

7.12 Section through rooflight at the Dulwich Gallery, by Sir John Soane.

7.13 The Sutton Hoo Museum and Visitors' Centre, by van Heyningen and Haward.

7.14 Sainsbury's Supermarket, Greenwich, by Chetwood Associates.

central space, such as in the Reform Club and the Oxford Museum as we have seen. This allowed buildings to become more than two rooms deep, but also brought the light of day into them. It is a particular pleasure to walk among the exhibits of the Oxford Museum, completely enclosed by the cloister walls, yet still be able to sense the rise and fall of the day, and changing weather as it makes its mark in the sky above. Since it was built before the invention of the electric light this made obvious good sense as it does from a sustainability point of view today, because it saves on energy used for artificial lighting. A top-lit atrium has become one approach frequently adopted for contemporary, deep-plan office buildings, and a variation of this was used at the Heelis Central Office for the National Trust, as we saw in the last chapter.

More typically, modern buildings that incorporate extensive daylighting from above use internal louvres to help control solar gain. One interesting example from the contemporary world of shopping is the Sainsbury supermarket at Greenwich by Chetwood Associates (Fig. 7.14). The building uses north-facing saw-toothed rooflights incorporating internal louvres to provide natural daylighting for general background lighting as part of a larger 'green' strategy. Supermarkets are often single storey 'sheds', so it is disappointing that so few consider using rooflights as here. At Greenwich artificial lighting is used only for displays, at check-outs and during the hours of darkness rather than its more ubiquitous use elsewhere.

Light at the edge

As we said earlier in this chapter (and have just seen at Sainsburys in Greenwich), a primary objective of sustainable design is to maximise the quantity and distribution of daylight into an interior – subject to concerns for glare and solar gain. Before he became preoccupied with the reciprocity between light and shadow, Le Corbusier had advocated rational means to achieve this. However, he rarely used rooflights, concentrating instead on increasing the wall's capacity to admit light. We saw this at the Salvation Army building where he proposed an all-glass façade or 'window wall'. Developing the logic inherent in this desire to maximise light led him to say that 'the window is made for lighting, *not for ventilation*'. The environmental discomfort that resulted from an all-glass façade played a part in his invention of the *brises-soleil.*

How best to maximise daylight was discussed in his book *Precisions* published in 1930, where he stated a fundamental principle to be that 'architecture consists of lighted floors'.[18] This was further advocacy for the practical advantage of strip windows over holes-in-the-wall for increased illumination. He demonstrates this by referring to a diagram from a photographic manual – a camera is a kind of light recording machine (Fig. 7.15). Exposure instructions contrasted a full-width horizontal window – the type made possible by Corb's 'Five Points' – with traditional hole-in-the-wall windows. According to Le Corbusier and his pseudo science, the strip window has two zones of bright light – 'very well lit' and 'well lit'. Conventional windows, in contrast, have four zones of rapidly diminishing light: 'zone 1 very well lit (two very small sectors), zone 2 well lit (a small sector), zone 3 poorly lit (a big sector), zone 4 dim (very big sector). The table adds, "you should expose the photographic plate four times less in the first room".'[19] This quasi-scientific demonstration of maximising access to light comes in a chapter entitled 'Techniques are the very basis of Poetry'. (We will see a variation of the singularly rational approach to maximise the quantity of light in Chapter 8.)

In earlier chapters we mentioned that Le Corbusier's design philosophy changed radically around the time of the Second World War. In particular, his attitude to light changed from maximising the quantity to a more poetic contrast of light and shade. Nowhere is his exploration of the reciprocity between light and dark more dramatically demonstrated than in his pilgrimage church of Notre Dame du Haut at Ronchamp. Coming closer to the church, which sits on a hill, the general level of light increases as the visitor leaves the wooded valley below. The inclined, white-painted wall combines with the famous soaring roof to act as a kind of concave lens, intensifying the

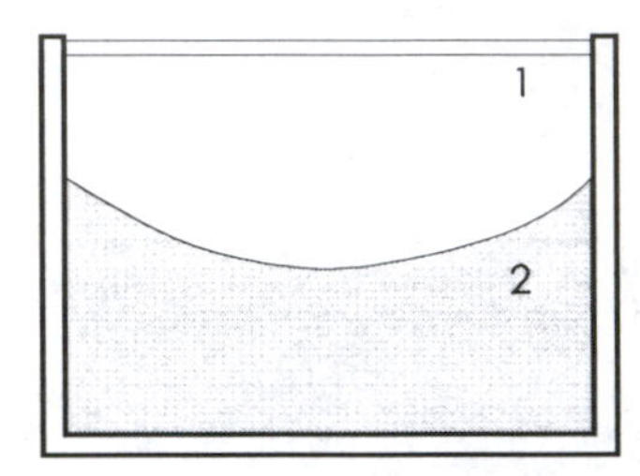

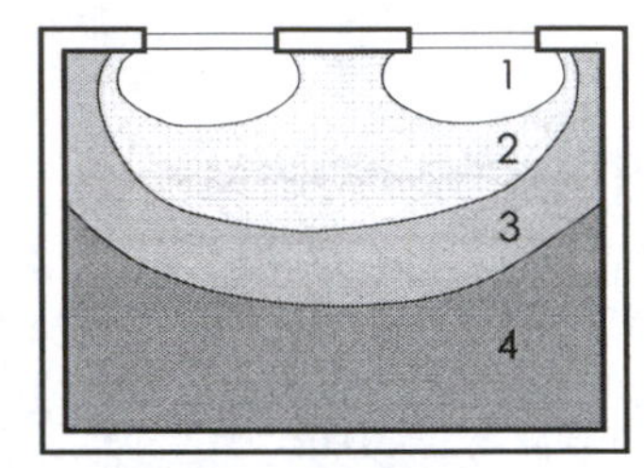

7.15 Diagrammatic plans showing the greater quantity of light that can spread across a room by using a strip window. (After Le Corbusier's sketch in Precisions.)

7.16 Notre Dame du Haut, Ronchamp, by Le Corbusier.

light in the approaching visitor's eyes (Fig. 7.16). On entering, the visitor is plunged into an unexpectedly dark space. The cavernous chapel is lit by a glow of coloured light from small windows in a thick, south-facing wall and a thin, horizontal strip of light separating wall from roof (Fig. 7.17). Instead of a bright space filled with light from on high as we would encounter in a Renaissance church, for example, we are immersed in darkness, like the seed, with a horizon of light high above us. This tactic of inverting tradition Le Corbusier used to distance the Modern Movement from history. Intense

7.17 Interior of Notre Dame du Haut, Ronchamp.

and dramatic light forcing its way into the unexpected darkness of the chapel invites the visitor to reflect upon the fundamental meaning of the sacred, which has been shown to underlie many different religions.[20]

Light had been the shaper of those most magnificent expressions of Christian architecture, the Gothic cathedrals. But the meaning and use of light was very different then. Light was considered the least material substance hence closest to the mysterious pure form of the Creator. God's presence was everywhere, God was light, hence light came to be considered the principle of highest value. According to the Book of Revelation, heaven itself was of 'pure gold, like to clear glass', and the Gothic cathedral came to be constructed in the image of the Celestial City. The development of the stained glass nave wall and associated flying buttresses was less concerned with increasing the quantity of light – as argued by Steven Jay Gould – than making light appear to be the material substance of the wall itself (Fig. 7.18).[21] Stained glass seems to hold the light in the wall rather than simply letting it pass through. From the inside, Gothic cathedrals seem to glow with the luminous nature of light, particularly those that retain their original thick medieval stained glass, as does Chartres.

The light level measured quantitatively in a Gothic cathedral with its original medieval glass is probably not that different from a medieval dungeon, but the atmosphere is very different. The way light is dealt with in the design greatly affects the mood of a space. Light forcing its way in through a small window feels very different from light held in the wall. The light level at Ronchamp is probably similar, but the particular way Le Corbusier designs the light to penetrate through many small but specific openings gives a mysterious aura and fresh meaning to that sacred space.

Orientation is an important consideration for lighting using wall openings. There are symbolic orientations and functional ones. Christian churches, for example, are oriented to the east so that the experience of entering and participating in this sacred space is a journey towards the rising light of day, the building and its use of light embodying the metaphor of Christ the light of the world. One nave wall is consequently on the south, the strong sunlight intensifying the glowing wall of a Gothic cathedral. Le Corbusier drew upon this fact in orientating the pierced wall at Ronchamp to the south.

7.18 Typical Gothic cathedral nave wall.

More functional considerations are involved in the quantity of light in relation to solar gain, which may suggest orientating larger windows to the north. Painters' studios often have north-facing windows not only to allow more light but also to ensure that its colour value is relatively unaffected by the sun's direct rays. We saw other functional considerations at Melsetter House, such as the preferred orientation of bedroom windows to the east, which catch the rising sun as people wake.

Soane's Dulwich Gallery contrasts the clear light and high level of illumination for viewing works of art with a melancholy, yellowish, low level of light for the benefactor's mausoleum located on the west of the building. The meaning of the spaces associated with such different functions is prompted by the mood made by the two very different kinds of light, westerly light carrying an element of melancholy as the day itself fades and dies. Asplund's County Courtroom in Solvesborg (1917–21) has a central rooflight over the court and one window facing west, the cool light of reason contrasted with the searching light of the afternoon sun, which was the time sessional judges made their judgement.

The architect/engineer Auguste Perret (for whom the young Pierre Jeanneret, later Le Corbusier, worked) made a modern interpretation of Gothic in his design for the church of Notre Dame du Raincy (1923–24). In place of stained glass windows, he designed the whole wall to the nave and chancel to be enclosed in coloured glass blocks, which 'hold' light in the wall in a similar way – although the light level is probably higher than its medieval predecessor (Fig. 7.19). Because it uses reinforced concrete, the structural frame at Le Raincy is more slender than a Gothic cathedral and the whole envelope is a shimmering wall of coloured light.

7.19 Church of Notre Dame du Raincy, by Auguste Perret.

From the mid-nineteenth century light and air became associated with the search for healthier living conditions in cities transformed by industrial production and population growth. From William Blake's 'dark and satanic mills' to James Thomson's 'City of Dreadful Night', darkness was a common perception of cities riddled with squalor and disease. Light became a metaphor for cleansing the unsanitary environmental conditions that prevailed. Glass blocks proved to be the perfect new medium for the expression of light in the architecture of the Modern Movement devoted to the ideal of purification. Glass blocks are a quintessentially modern material. Building blocks associated with industrial production, they seem to combine the practical with the poetic. They are available in differing degrees of opacity, but all have an integral glowing luminance, which appealed to the Modern Movement's linking of light with health.

This new building product was used by Berthold Lubetkin (who had worked in Paris) and Tecton for the front façade of the Finsbury Health Centre (1938). A central part of their 'Finsbury Plan' was to bring light and air into a slum-ridden area of London, and the Health Centre's principal façade is a glass block wall through which patients enter (Fig. 7.20). The entire south-facing wall fills the inside space with a high level of illumination, light fills the wall itself, and its opacity provides privacy to the patients in the waiting area.

Developing the edge

Glass blocks provided poetic moments to what was largely a rational or functionalist Modern Movement that allowed for little influence on a building's design other than the programme, the site and materials – which had to be modern. There was little explicit consideration of the human spirit itself. Modernists believed that sticking to the slogan 'form follows function' would in itself produce beauty, consequently mankind's spiritual and other needs would be satisfied by the purely rational concerns of planning and construction.

Amongst post-Second World War architects much influenced by Le Corbusier, Louis Kahn developed a conception of architecture that was founded on light and

7.20 Finsbury Health Centre, by Berthold Lubetkin.

addressed the needs of the human spirit directly. For Kahn, 'Silence and Light' was a metaphor that he used to explain his understanding of Creation and also the aim of artistic creation itself. Silence represented that which does not yet exist, the unmeasurable. 'I sense Light as the giver of all presence's, and material as spent Light,' he wrote. 'What is made from Light casts a shadow, and the shadow belongs to the Light.'[22]

Although Kahn's notion was poetic and metaphysical in intent, his conception of light does have some basis in physical reality. The sun that gives us light also provides the energy needed to set in train the processes of growth from which springs life and that ultimately gives us some of the materials from which buildings are constructed. Drawing up moisture, the sun plays a part in producing clouds from which falls the rain that along with wind and frost wear away the Earth's surface and weathers buildings.

The role of art or architecture, in Kahn's conception, is to make light visible in some material way. These allusive notions he gave concrete form and they became the poetic basis for his design intentions for specific buildings such as a library. 'I see the library as a place where the librarian can lay out the books, open to selected pages to seduce the reader ... and the reader should be able to take the book and go to the light.'[23] Kahn's library at the Philips Exeter Academy, New Hampshire (1967–72) puts this idea into practice. At the centre of this great, four-square, fortress of a building is a triple-height space where individual books can be displayed on tables and also the collection can be seen through large circular openings that surround the entrance hall. Top lit by clerestory windows, the deep concrete beams spanning this space deflect light of an even quality down into it. Kahn described the building as 'a concrete doughnut where the books are stored away from the light'.[24] The section shows the organisation of the building with the book-stacks in the middle zone out of reach from direct sunlight (Fig. 7.21). Yet light can spill into this zone from both the centre of the 'ring doughnut' and the tall windows.

Built into the external wall are timber study carrels where the reader can sit by the window. Each reader has a wooden shutter to control the amount of light or the view. The edge zone of the library is double-height with a large area of glass above the carrels that allows light to penetrate across the section and illuminate the book-stacks. The section is cleverly worked here with mezzanines that cut off sunlight at the same

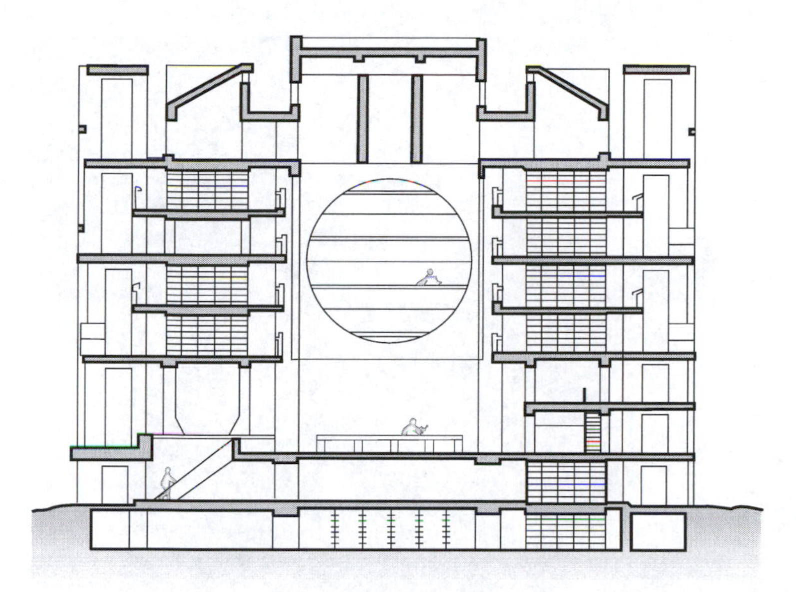

7.21 Section through the Philips Exeter Academy Library, New Hampshire, by Louis Kahn.

time as they make the top bookshelves accessible. The carrels are expressed on the exterior as human scale timber elements within large glass-filled openings.

The Exeter Library combines a particularly thoughtful use of materials in response to how humans use space and feel comfortable in it. Daylight is also introduced in a modulated way that responds to three realms of use: filtered top-light of a general and modest level for the central circulation space; shaded but well-lit space for book-stacks; and intense but controllable light for reading. The section is his main tool for working out the arrangement for distributing the appropriate quantity and quality of light, and also to give meaning to the building for its users.

A unique use of the glass wall and the edge condition characterises Peter Zumthor's Art Museum at Bregenz on Lake Constance in Austria. Zumthor's aim was to make the building have something of the quality of light that is characteristic of the adjacent lake and the misty air that slides across it (Fig. 7.22). What appears to be an opalescent box of light is not made from glass blocks, nor a conventional glass curtain wall, but from large sheets of overlapping glass rather like tile-hanging on a gigantic scale. The slight raking angle of these opaque sheets subtly increases the amount of reflected light that strikes the eye. In the day a pearly luminesence has something of the quality of light on the adjacent lake, and at night it glows softly from within. Daylight is funnelled deep into the centre of the galleries by what Zumthor calls a 'light plenum', effectively an opaque glass suspended ceiling with sufficient void above for light to spread across. Louvres within the wall sandwich give control over the amount of light that can penetrate, and gaps between the sheets of glass allow fresh air to enter. The building is planned such that all movement from floor to floor entails going to the edge, to the light and the air.

In general what one often finds is that the building envelope creates a condition of contrast. One is either inside or outside in a similar way to either being inside or outside of a site boundary or the walls of a medieval city or the wall of a cell. This

7.22 The Bregenz Art Museum, by Peter Zumthor.

condition has both emotional connotations such as feelings of inclusion and security and functional implications. For example, as we have seen in Chapter 6, ventilation needs lead to the stomata of leaves and air inlets in buildings. The need for light has given us a vast vocabulary of window types, glazing materials, solar shades and a variety of reflectors. These last can be either external or internal. In the narrow streets and lanes of Cambridge, England of pre-Industrial Revolution times, white-painted hinged panels fixed outside at the cills of windows were used to direct light into rooms. More recently light shelves fitted inside the upper third or so of windows reflect light on to the ceiling and then into the space to create a more even distribution of light. This, of course, reduces the degree of contrast at the edge. Thus we have two sustainable solutions, one derived of necessity from a time when energy was scarce and another developed in response to our present need to make the best possible use of natural daylight to reduce energy consumption.

Light and place

In his *Critical Regionalism*, Frampton has recommended specific consideration of 'the range and quality of local light' as one strategy for resisting the universalising tendencies of globalisation that manifest themselves in architecture today.[25] He criticises the exclusive use of artificial light in many art galleries, for example, because 'this encapsulation tends to reduce the artwork to a commodity'. He argues instead for a more open relationship between inside and outside, but using monitors, blinds and screens to prevent intrusive sunlight – as we have seen at Sutton Hoo. This approach to designing with daylight would 'guarantee the appearance of a place-conscious poetic in a form of filtration compounded out of an interaction between culture and nature, between art and light'.[26] Frampton's *Critical Regionalism* chimes with the argument advanced here for the need to consider the design of daylight in buildings as a vital part of a more sustainable approach to the environment, in the broadest or deepest sense.

In Chapter 2, we discussed the importance of temperature in creating particular habitats to which creatures adapted and to which vernacular buildings responded. The enormous difference between daylight in southern and northern countries and how this has found a response in art has been pointed out by Norberg-Shultz. Renaissance domes he considers to reflect the Mediterranean sense of a clear sky that appears as a luminous vault spreading expansively from horizon to horizon. The dome in churches lit from the apex represents this idea of lucid heavenly light. In contrast, Nordic light is more often seen veiled by cloud and with its source lower down near the horizon.[27] Scandinavian architects such as Alvar Aalto make frequent use of light striking obliquely across the section from a clerestory window rather than from a skylight. The library at Mackintosh's Glasgow School of Art also demonstrates this approach, light coming in this case from one tall window and filtering its way across the space through structural posts and attenuated vertical balustrades rather as it filters through winter trees in a northern forest (Fig. 7.23). Some architects of the best

7.23 The library at the Glasgow School of Art, by Charles Rennie Mackintosh.

buildings have used the characteristic light of their region to create spaces that reflect their locality, hence its quality reinforcing a sense of belonging. This contrasts with the more universal approach of Miesian glass-walled buildings in any place in the world, designed independently of conditions.

Present and future

The question of natural illumination in buildings is perhaps the clearest example of how the quantifiable and the qualitative need to be intertwined in design. As we have suggested, the optimum quantity can be calculated for many activities, but the human experience of the activity can be enhanced by the kind of poetic considerations outlined above. Often this is achieved by the appropriate contrast of light and shade.

The future of light and shade in sustainable architecture will be an art of balance. Just as plants themselves need to operate in extremely variable conditions, and do so by photosynthesising very efficiently at low levels of light, but in full sunlight reject potentially harmful light by dissipating it as heat,[28] so we will find the right balance for our buildings and us. This is likely to be a regional approach where the mix of factors is different. In cloudy northern Europe we need to design for high daylight factors (insulating the glazed areas at night to reduce heat loss) but control high solar gains which are less common. In the sunny Mediterranean region we need to pay more attention to the abundant sunshine (for otherwise cooling loads will be excessive). So depending on the climate of the area, the balance will vary, and this creates opportunities for different forms, fenestration and poetic approaches. We may see a specifically environmental architectural language develop that has regional dialects. (And similar considerations will apply in developing a language of urban design in relation to the grain of existing cities.)

There are considerations of morphology, pattern, context and technology. Glazing is, after all, just another form of light collector, and one sees both man-made examples and natural ones, such as leaves as one strolls through any city (Fig. 7.24). In the last chapter with the Heelis National Trust offices we saw an example of large areas of fixed glazing to bring light into a very deep plan. It is also, of course, possible to use more mobile, responsive systems which adjust quickly to the environment.

Traditionally, particularly in Mediterranean climates, shading systems that could be put into various positions and also allowed ventilation were common. At the BRE Environmental Building adjustable louvres are used to control solar gain and glare. Through a combination of advanced materials and control systems we are likely to develop ways of using the sun's energy for lighting or heating when and only when we want to. This greater adaptiveness, of course, has numerous counterparts in the biological world – for example, the iris of the eye controls the size of the pupil and thus how much light reaches the retina.

What is important is that these new developments will affect the skin and periphery of buildings and might have something of the elegance of their eighteenth-century Georgian forebears, which evolved from vernacular traditions by adapting to new social demands of refinement and classical ideas of proportion and balance. Our own cultural expectations may be rather different, but architecture will benefit by being similarly brought into harmony with inherited traditions of building that reflect a deeper structure to life, or the way we live, at the same time as adapting them to the new possibilities presented by technology.

7.24 Man-made and natural light collectors in London.

Building references

Sutton Hoo Museum and Visitors Centre: Architects – van Heyningen and Haward; Environmental Engineers – Max Fordham; Structural Engineers – Price and Myers.

Sainsbury's Supermarket, Greenwich: Architects – Chetwood Associates; Environmental Engineers – Buro Happold; Max Fordham; Structural Engineers – WSP.

8

Cities

Introduction

The pleasures of urban life have been known for millennia and the very word 'civilised' associates higher standards of comportment and learning with the greater densities and diversity of city living. The word 'culture' (which is almost synonymous with civilisation) originates in the Latin word *colere*, to cultivate; *cultiva terra* was medieval Latin for 'arable land'. The words 'urban' and 'urbane' (which is defined as 'having the manners or culture characteristic of town life') are derived from the Latin *urbs*, which originally meant the curve of a ploughshare.[1] The origins of these words – civilisation, cultured, urbane – reminds us how the aspirations of humanity began with an attendance upon nature, rather than a separation from it. As we have seen, some cultural artefacts were designed to mediate between the human and natural worlds. Cities may play a similar role in the future. Our current pressing need to live more sustainably, combined with the fact that the majority of the world's population lives in cities, makes it timely to reflect upon the relationships between the city, the region and the natural world.

Raymond Williams has described how 'culture' became a term critical of civilisation, cities and industrialisation at the end of the eighteenth century. Rather than in theatres, clubs or coffee houses, the Romantic poets immersed themselves in remotest nature to find the highest truth. Rousseau's 'Noble Savage' was only the most extreme form of this widespread conception. The country became more and more contrasted with the city as industrialisation left its mark on Victorian life but people were drawn to the city to find work. They had to live there until steam trains could transport them to the suburbs and their cottage-like homes. The retreat from the city centre at the end of the nineteenth century was followed early in the twentieth by radical proposals for a new form of city. In response to the dark, dirty, disease-ridden Victorian city of streets lined with buildings, the Modern Movement proposed cities of tall, isolated blocks surrounded by open green spaces full of light and air.

Daylight, fresh air for natural ventilation and the inclusion of plant and animal life in the built environment are, of course, some of the goals of sustainable design. But the city, which is associated with the accomplishments of civilisation, is at times referred to today as a 'concrete jungle'. The physical solutions proposed by Modernism have produced their own psychological problems. The modern metropolis has become 'home' to most of the world and so it should meet not only our physical and economic requirements but also our emotional needs and desires. To sustain life in the fullest sense, the city of tomorrow must consider not only the physical dimensions of a healthy and sustainable environment but also what has been called mankind's 'existential security'.[2] One of the challenges of a sustainable city is how to 'repair' the systems we have and develop new ones so that we can live more 'naturally'. This chapter briefly sketches some of the considerations involved in doing so.

In the final chapter of his *Court and Garden*, Michael Dennis describes what he calls 'two traditions': 'the pre-modern (would pre-Enlightenment be more accurate?) tradition ... composed of articulated spaces', and the modern 'tower city' that 'favours rationalised and articulate solids instead of contiguous ones'. His conclusion that these 'two traditions constitute a rich legacy, not a confusing obstacle'[3] chimes with the argument advanced in this book. It suggests that a careful balance of principles drawn from a study of mankind's aspirations as embodied in existing cities alongside empirical objectives could form one of the foundations for sustainable cities of today and tomorrow.

City and region

One way of viewing cities is as networks of buildings and spaces, which have accumulated meaning over a period of anything from decades to many centuries. They have been intimately connected to their regions and thrived on cyclic processes. Traditional Chinese cities, for example, disposed of their 'night waste' on the fields around the city, and the fields in turn provided food. Similarly, in a burgeoning nineteenth-century Paris, transport and food were linked. Manure from horses, which pulled everything from carts to people in public transport carriages, was removed and used as fertilizer for the extensive market gardens serving the city. Parisian homes and restaurants transformed the produce from the hinterland into potato salads, omelettes with asparagus, and dozens of other dishes celebrated by Proust and other writers.

Before the Industrial Revolution there was almost always an integral relationship between city and country – in some cases this continued into the nineteenth century (for example Paris, as described above) and even the twentieth century (as noted below). The city drew most of its supplies from the immediately surrounding hinterland and in turn it was accessible to people living there for special needs – markets, defence, politics, religious festivals, etc. Industrial production created a demand for more raw materials and for energy to drive the machines. Technology in turn created the means to move material and goods further and faster from the cities.

The invention of railways followed by the car also allowed people to live ever further from the centre of cities. This has led to cities of enormous populations expanding over very large areas of land. In 1800, for example, London was the only city in the world with one million inhabitants. Now there are hundreds of cities with that population or more and London's present population of approximately seven million is relatively small compared with many from around the globe. New York's population has grown only 5 per cent in the last 25 years, yet its surface area has increased by 61 per cent.[4] Increasingly fast transport and extensive networks of communication have broken the interdependence of cities and their hinterland.

As city populations grew they demanded more and more resources while at the same time their buildings and roads sprawled outwards consuming the surrounding land. Cities have become the major centres of resource consumption – meat and vegetables, wood and metals and a vast range of materials, consumer goods and energy. Cities have been compared to organisms in that they have a distinct metabolism[5] and some have become so big that they have been described as 'super-organisms'. Their dependency on an ever wider environment has led to their being called 'parasites'.[6] Biologically, what is required instead is a state of symbiosis in which cities both consume and produce in the most ecologically appropriate ways.

Writing in 1996, Girardet could say that most Chinese cities remained largely self-sufficient for food, drawing their supplies from adjacent farmland. In his Reith Lectures of the same year, Richard Rogers confirmed this, describing Shanghai, where he was working, as self-sufficient in vegetables and grain.[7] But his experience was one of a city being rapidly developed along Western lines of planning, dominated by zoning and fast roads, a system devouring land and contributing to terrible pollution. Rogers' solution to the problems caused by energy-consuming urban sprawl is the 'Compact City'. A combination of more compact cities, replacing a linear, mechanistic economic model with a cyclical, biological one and using fewer resources is a way to make more sustainable cities.

Our view of some of the considerations involved in a more biological and cultural approach to urban design is outlined below.

A short history of urban form

A sustainable city will need to consider more than the quantifiable data of environmental design and advanced technology. The wider socio-economic factors and the less quantitative aspects of our lives need to be taken into account. The pattern of cities as found and operating reasonably well in some areas provides a useful counter to the generalised conceptions of modern planning, which favours zoning, separation of activities and the car. In the earlier chapters on Site and Setting and Building Design we outlined a history of how humans have occupied the Earth in order to show how traditions of human habitats could inform sustainable design today. In a similar way, this section will start to explain why a balance of the traditional city matrix of spaces and modern technology offers a way of making cities more sustainable.

Cities emerged in six widely separated localities around the world at different times – Mesopotamia, India, Egypt, China, Central America and Peru.[8] Lying between the rivers Tigris and Euphrates, Mesopotamia in modern Iraq was the first 'urbanised' society. The trend to urbanisation began there about 7,000 years ago. Within 1,000 years there were many cities in the region and by 4,000 years ago it is estimated that about 90 per cent of the population lived in cities. Some were quite small by today's standards but Babylon was of considerable size by about 2000 BC. Population figures are hard to establish, but a clay tablet from the region gives an indication of how big cities had become, recording an order for 150,000 bunches of onions.[9]

One theory suggests that itinerant craftsmen trading in specialised commodities allowed the expansion of what had earlier been little more than large villages, like Çatal Hűyűk, in Anatolia.[10] Aerial archaeology has shown a vast network of canals irrigating some 17,000 km^2 of the 'Fertile Crescent', hence capable of supporting large cities such as Babylon and Ur. We also get a glimpse of the inter-relatedness of city and region, and how this allowed civilisation to flourish.

Investigations at Ur – one of the historic capitals of ancient Sumeria – indicate that cities grew organically in 'a natural unplanned process whereby an urban settlement evolves from a village',[11] rather than a predetermined plan. This region is extremely hot – for temperatures to often exceed 40°C and 50°C is not unknown. Courtyard houses

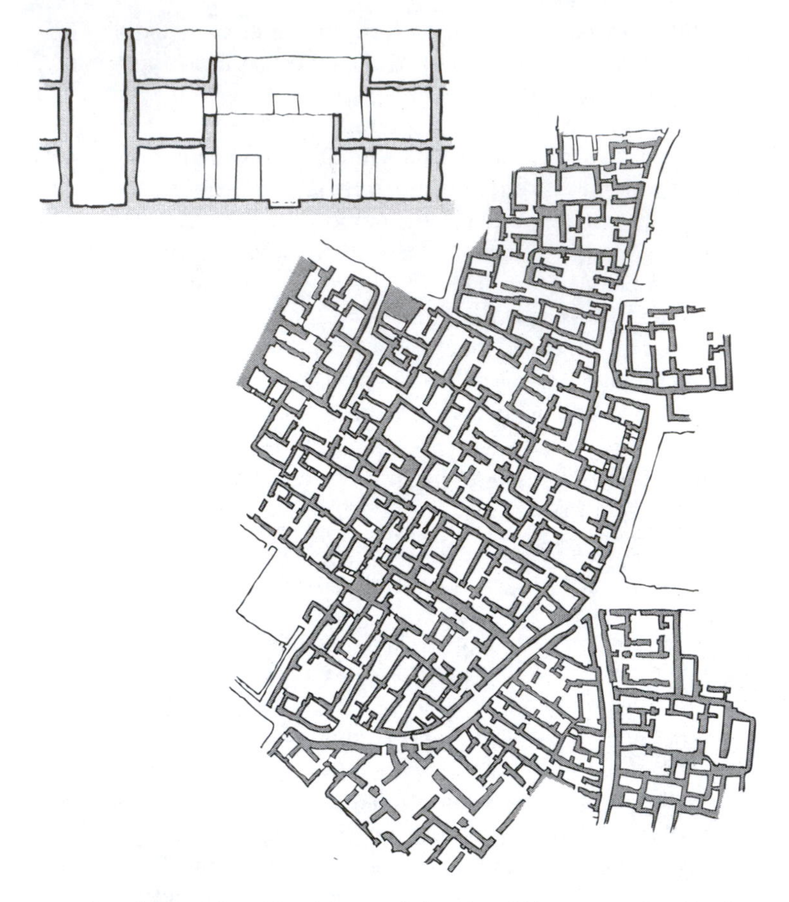

8.1 Detail plan of residential part of the city of Ur as excavated by Leonard Wooley.

of the type described in Chapter 3 and shown in Figure 3.3 formed the basis of urban settlement, tall and narrow to exclude the sun (Fig. 8.1).[12]

Direct links are known to have existed among the Sumerian, Egyptian and Greek civilisations and cities of all these regions were composed of a matrix of narrow streets and courtyard houses, with the addition of citadels and temple precincts. In Greek city states (750–500 BC), the role of public space became particularly important with the 'agora' accommodating social life, business and politics becoming the heart of the city.[13] In ancient Greece we also find the first systematically planned cities, such as Miletus and Priene, laid out on a modern-looking grid (Fig. 8.2). The first recorded town planner emerges at this time, Hippodamus of Miletus setting out the reconstruction of the city in 479 BC after its destruction by the Persian army.[14]

The grid had a metaphysical meaning in ancient cultures which is very different from the purely pragmatic organisation of land use in its modern counterpart. The Greeks likened their city grids to weaving. Weaving brings something into being. A Mesopotamian myth described the creation of the Earth from a woven reed mat. The god Marduk 'planted a wicker hurdle on the surface of the waters. He created dust and spread it on the earth'.[15] This idea that the Earth was derived from weaving probably informed the Greek view of city planning. An Athenian oath taken by its citizens

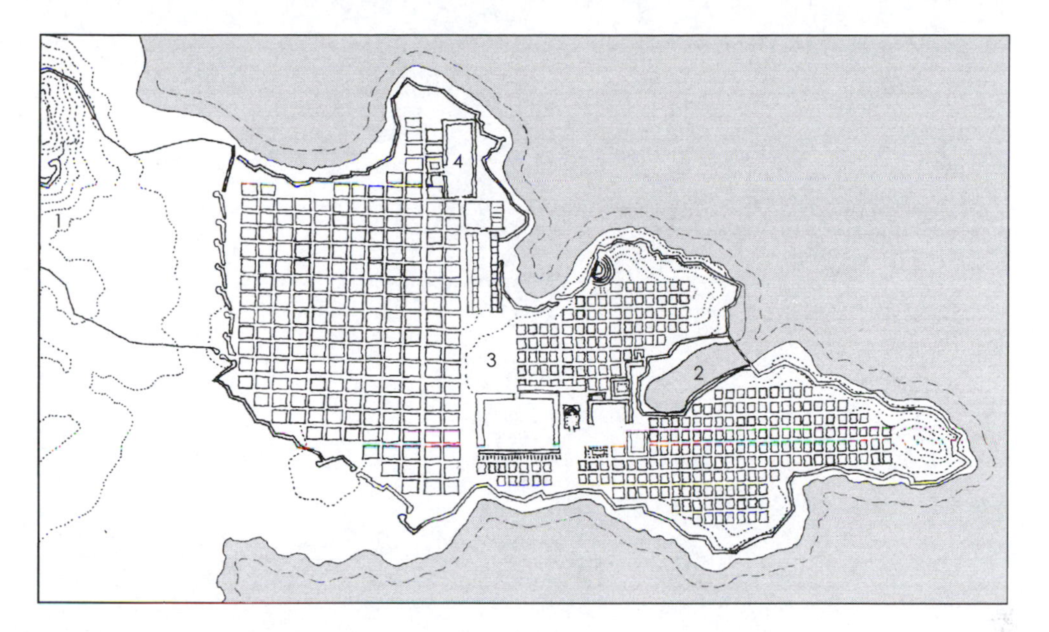

1. Early fortified hilltop settlement
2. The main harbour
3. The Agora complex
4. Theatre and other cultural/leisure activity facilities

8.2 Plan of Miletus.

was 'that we will transmit this city not less, but greater, better and more beautiful than it was transmitted to us'.[16] This sentiment was echoed in the Kyoto Agreement in relation to the Earth's resources and would serve well as a guiding principle for us today.

The Romans, who took over and adapted so much of Greek culture, also laid out their cities on a grid, but one that was organised around two dominant streets perpendicular to one another that crossed at the centre, the Forum (adapting and extending the role of the agora). The *cardo* ran north–south, the *decumanus* ran east–west (Fig. 8.3). It has been said of the Romans that they were orientated to the cosmos in their cities because of this organisation. They always knew that they followed the sun's daily path and were going towards the sun, or towards that part of the Earth where the sun never went.[17] Like their Greek predecessors, Roman cities were composed of narrow streets with courtyard houses on either side. But in large metropolitan cities such as Rome itself apartment blocks emerged and some streets were lined with shops, their fronts little different from those in the smaller streets of today (although they would not have been glazed; see Fig. 8.4).

Medieval European cities, developed in part in response to a need to provide security, tended to have much less ordered patterns. Streets were often narrow and access to light and air restricted, but these cities, particularly with modern drains installed and traffic controlled, remain among our favourite 'romantic' weekend breaks. Their apparent disorder contrasts with the 'classical' city forms. These instead emphasise order, rationality and clarity (as well as cosmological connections), and are often based on a grid as we have just seen in Greek and Roman cities. In China the great urban centres of the Tang Period, AD 618–907, such as Chang'an (the capital)

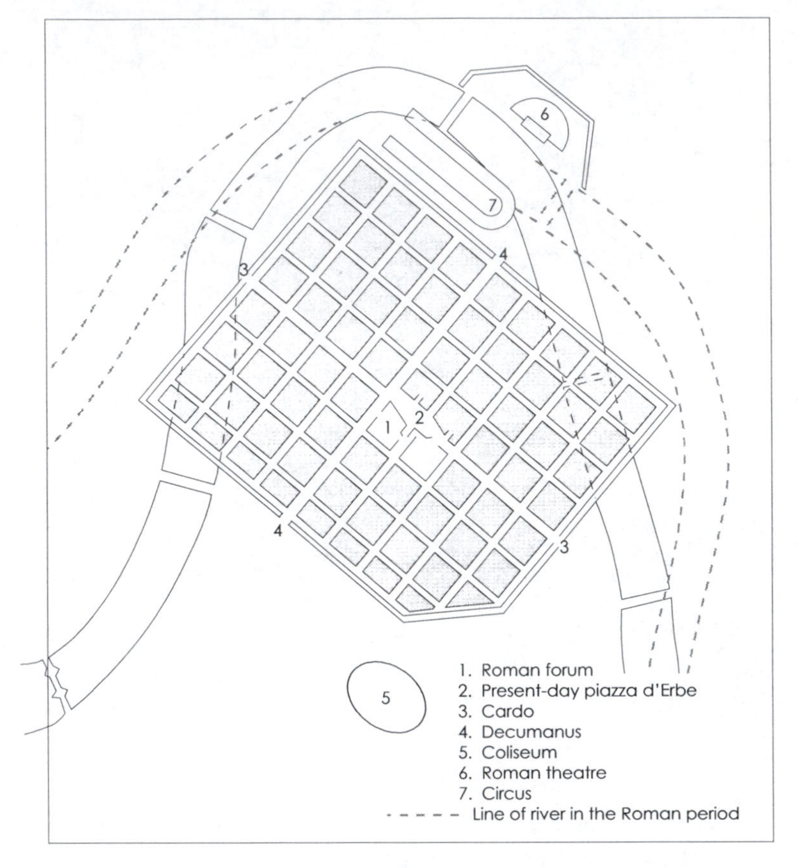

8.3 Plan of Roman Verona.

8.4 Roman street leading to Trajan's Market in Rome.

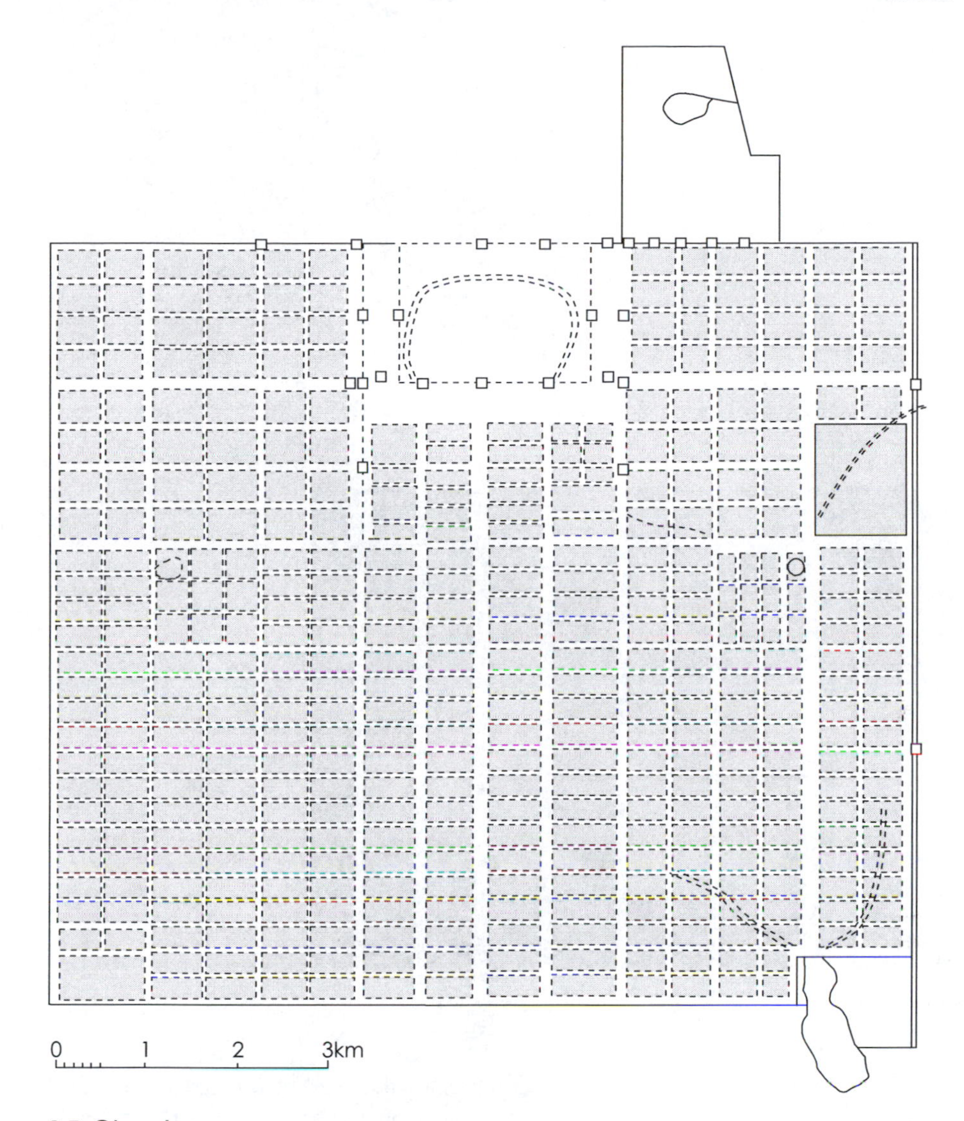

8.5 Chang'an.

with over a million inhabitants were laid out with rectangular grids of parallel streets and boulevards (Fig. 8.5). And the admirative Japanese copied this for their capital at Nara in 694 and later in 794 at Heian (now Kyoto). These cities were set out according to the Chinese art of Feng Shui which incorporates principles relating to the points of the compass similar to that of Roman planning.[18]

Another aspect of the development of medieval towns and cities was their organic, unplanned growth to serve their primary role as markets within the city walls (if they existed). Their patterns often have a serendipitous feeling that one finds in the drawings of Paul Klee (Fig. 8.6).[19] One or more public open spaces might specialise as a market, but the 'entire medieval city was a market'.[20] Market spaces might be no more than a widened street, but all street frontages were important because shops and workshops dominated city activity. A medieval city is often characterised by a significant

8.6 Paul Klee's 'City of Cathedrals' (redrawn).

8.7 Medieval market place at Kingston-upon-Thames.

public space in front of the cathedral and another for the market presided over by a guildhall (Fig. 8.7). Narrow passages typically open off streets providing access to courtyards in back garden space. The City of London is a good example.[21] In England the university cities of Oxford and Cambridge adapted the monastery cloister courtyard as their model for developing urban spaces for this new activity. The richness of mixed use in compact medieval cities partly explains their attraction to our modern eyes.

As in the arts and architecture, Renaissance Italy set the pattern for developments in urban design in Europe from the fifteenth century onwards. The typical characteristics of Renaissance planning were the creation of new, usually geometrically defined and enclosed public spaces connected by a pattern of straight streets often having an axial relationship to significant buildings or monuments. A good example of Renaissance city planning is found at Nancy where a major expansion of the city was connected to the medieval core by a series of linked formal spaces (Fig. 8.8). In the Renaissance two significant changes appeared that were to have a major impact on twentieth-century town planning. One was the desire to embody utopian ideals – in

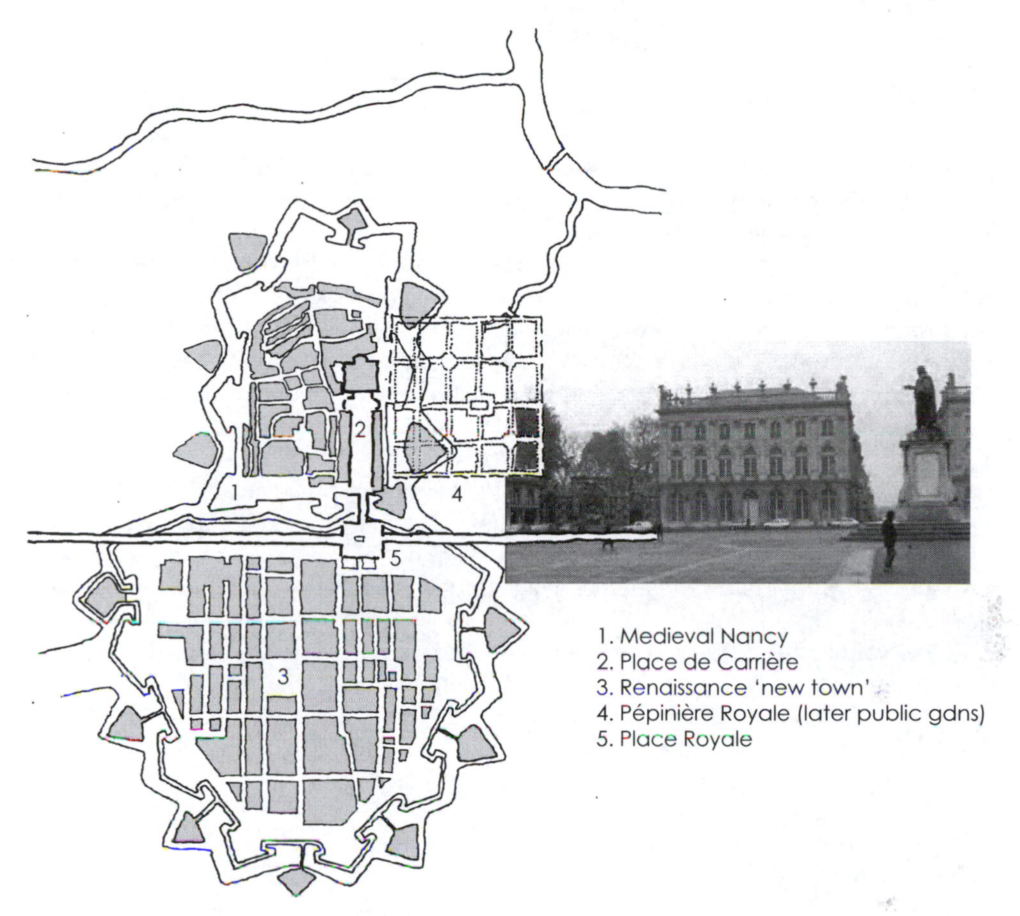

8.8 Plan of Nancy with photograph of the Place Royale.

some cases expressed in a city plan based on a perfect geometrical figure – and the other being the idea that the built environment should express 'theories concerning goodness, efficiency, well-being, etc.'.[22] We will return to these concerns in a moment.

The Renaissance tradition of formally organising public spaces in grand geometrical schemes developed through several phases but continued in a recognisable form onto the nineteenth century in Haussmann's Paris, for example. Like its predecessors, behind the classically co-ordinated façades of Parisian boulevards, a network of courtyards provided light, air and layered degrees of privacy and territorial control for its inhabitants. This arrangement is similar to the way courtyards were used to introduce daylight and natural ventilation into the deep plan of the National Trust Central Office discussed in Chapters 4 and 6 (Fig. 6.16).

The typical nineteenth-century Victorian city was largely the product of laissez-faire economics and the Industrial Revolution. This led to the dark and disease-ridden city that resulted from a hugely increased population living in inadequate, overcrowded accommodation or in gridiron rows of terrace housing with little open space. This City of Dreadful Night (as mentioned previously) led to the desire to escape to garden suburbs where the myth of the yeoman's cottage provided a dream of fulfilment and a new form of settlement. This desire for a house surrounded by its own land lingers on in the popularity of suburban housing developments today.

At the end of the nineteenth century Camillo Sitte wrote a book with the unlikely title, from a modern perspective, of *City Planning according to Artistic Principles*. He drew simplified plans of most of the outstanding city centres in Europe, demonstrating that their beauty stemmed not only from great monuments, art and architecture, but also from a combination of plazas and streets (Fig. 8.9). The spatial structure of the city was a matrix of bounded spaces that responded to the need of movement and to the proper placement of statues and architectural monuments. Sitte compared city squares to our living rooms in which we display our best things or the values that we aspire to. In the past, of course, many of these things have been sculptures of kings, nobles and leaders. But sometimes in fountains and pools we see the celebration of water, so vital to life in cities. War memorials remind us of more ordinary people and their sacrifice. But everyday objects such as telephone boxes can strike a pleasant note. In Barcelona's new Plaza del Mar a huge canopy of photovoltaics takes pride of place, providing shade at the same time as generating electricity. Perhaps the sustainable city of the future will make eloquent display of not only photovoltaic cells but wind turbines, rainwater stores and methane digesters imaginatively sited and appropriately scaled.

A few years after Sitte's attempt to recover the best from history, a planning movement arose that totally rejected the past. This found clear expression in Le Corbusier's *The City of Tomorrow*. The first chapter is called 'The Pack-Donkey's Way and Man's Way', in which he characterised European cities as finding their spatial arrangement

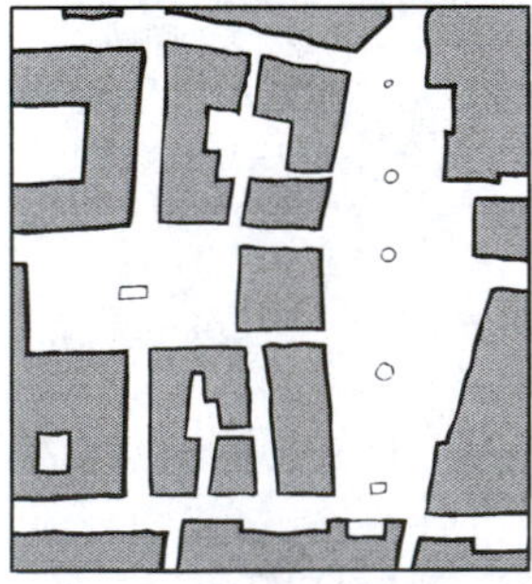

8.9 Plan of Piazza d'Erbe and Piazza dei Signori, Verona, with statue of Dante.

by the meandering path and the organic growth of markets. In contrast, he says, 'Man walks in a straight line because he has a goal and knows where he is going.'[23] Le Corbusier was much influenced by the 'Neues Sachlichkeit' (New Objectivity) movement that rose to prominence in the 1930 Congres Internationaux d'Architecture Moderne (CIAM), whose theme was 'Rational methods of site planning'.[24]

CIAM produced a purely functional analysis of the city. The four functions of the city – work, housing, recreation and transport – were to be zoned separately with fast road circulation between the zones the key to its efficiency. This became the basis of subsequent town planning policy and this over-simplification of the city has remained entrenched (although there are signs of its imminent demise as the dominance of the car, the ultimate twentieth-century machine, is called into question). Traditional arrangements whereby buildings lined both sides of streets or were grouped around squares were to be replaced by purely rational principles derived from the optimum orientation for sunlight with simple, rectangular blocks of building running in parallel, their height and spacing determined by sun angles. The approach of this movement is epitomised in a diagrammatic section drawn by Walter Gropius (Fig. 8.10). The

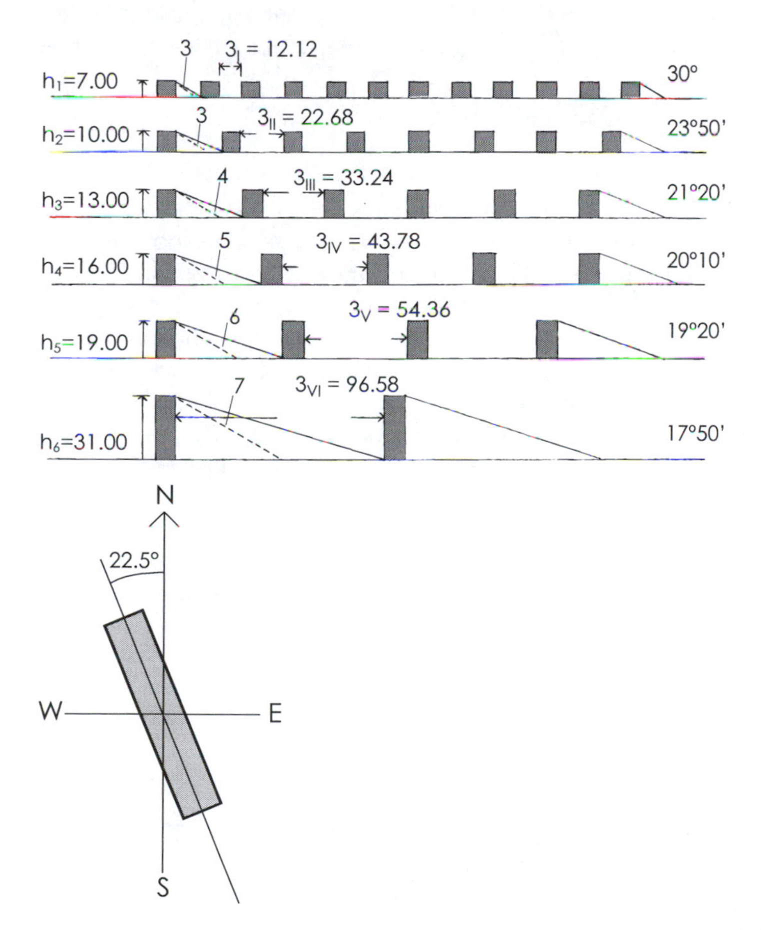

8.10 Diagram of optimum distance between blocks to obtain sunlight and the advantage of high-rise by Walter Gropius (above). Diagram of optimum orientation of dwellings by Walter Schwagenscheidt.

application of a singular determinism using scientific principles and a tabula rasa approach, rejecting tradition or history in favour of abstraction and machine-like imagery, has contributed much to the failures of the modern built environment. It perhaps gives the most sun (and ventilation) but little else.

The early years of sustainable thinking were dominated by passive solar design that concentrated on mono-pitched buildings orientated on plan to maximise solar exposure. It soon became clear that this strategy could not make a recognisable urban environment.[25] Instead we need to weave together a number of strands simultaneously to create something richer and this requires a historical appreciation of what has been done before. Le Corbusier's Contemporary City plans of 1922, reflecting the Modern Movement's interest in sun, light, air and cars, were arranged on a rectilinear pattern (Fig. 8.11) with the buildings spaced out in a kind of urban park and trees looking like sheep in pens.[26]

Delightfully enough, he departed for a moment from this strict rationality and linearity after becoming enamoured of the (charming and curvaceous) American entertainer Josephine Baker who took Paris by storm in the 1920s. Figure 8.12 shows them at a costume party in 1929 – Le Corbusier is the man without a hat but with glasses on the left. However, by 1930 his plans for la Ville Radieuse (the Radiant City) found him back on grid.

The legacy of the modern city

Despite the noble intentions of the Modern Movement to clear away the dark and unhealthy conditions of the nineteenth-century city, modern site planning has produced many appalling environments, particularly in the case of post-Second World War housing estates. By simply considering buildings in relation to one or two quantifiable criteria, paying little or no attention to existing form and context and ignoring or destroying any sense of place where it existed, the broader questions of how human beings respond to spatial and environmental conditions were overlooked. This is an important lesson to be kept in mind when designing the sustainable city. We need to maintain something of the modernist appreciation of light and air while at the same time creating cities that make us happy, fulfilled and which have a minimal impact on

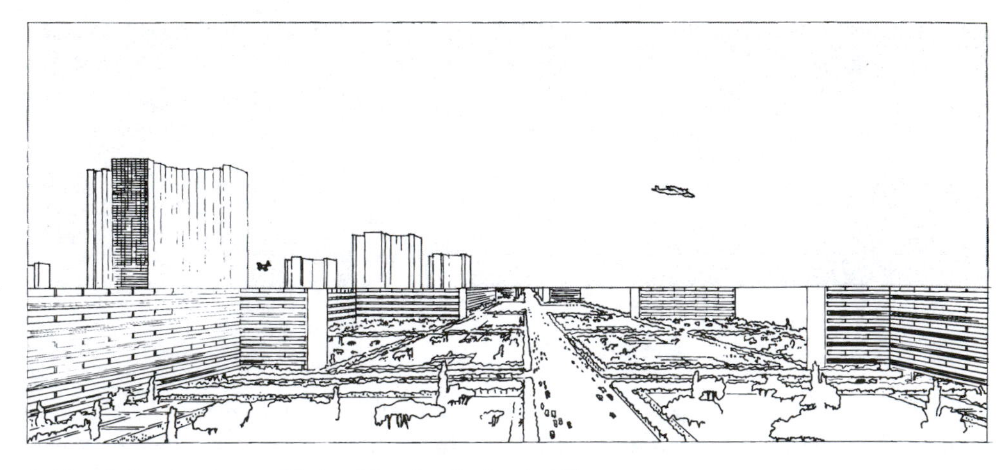

8.11 'A Contemporary City' – (after Le Corbusier).[27]

8.12 Le Corbusier, Josephine Baker and others.

the broader environment. Interestingly, recent research indicates that for people walking about cities, increasing diversity of built form and physical conditions (temperatures, light and shade, etc.) leads to greater satisfaction with their environmental comfort.[28] The literary model for this might be Joyce's *Ulysses* in which Dublin is seen as a tapestry of the noise of horses' hooves, the sounds of human voices, the smells of kippers frying and views of the city's power station to cite just a few of the strands.

Our view is that environmental considerations can inform the urban morphology (orientation, structure, street pattern, grain, landscape, massing, heights, etc.) but that they should not dominate. The potential of a site needs to be understood and a sense of place created. Aristotle said the city must be so designed as to make its people at once secure and happy. In order to realise this, as Sitte has written, city planning should not be merely a technical matter, but should in the truest and most elevated sense be an artistic enterprise.[29] In his book *Genius Loci*, Norberg-Shultz tried to show how the fabric and form of cities crystallised certain features of the surrounding landscape. In this way cities belong to a particular place and develop a distinctive sense of place themselves with their arrangements of streets and plazas, buildings and monuments, as described by Sitte.

The zoning and transportation model of planning, with its emphasis on the car, lent itself to the overriding goal of efficiency demanded of development in the second half of the twentieth century. This in turn led to the extensive disarticulated peripheries of business parks and shopping centres characteristic of our present cities. These single-function spaces linked to residential suburbs by fast roads drain the life from city centres which are reduced to commercial centres that are 'dead' after working hours. The sprawling city that developed in this way has become a major focus of environmental and ecological concern.

We mentioned earlier Michael Dennis' view that these two traditions of city planning – Sitte's pre-modern one of defined spaces and the Modern Movement one of rational free-standing blocks – form our legacy and that we must draw upon both. In *Court and Garden* he described seventeenth-century Parisian town houses where he emphasised how the defined spatial structure of forecourts mediates between the public realm and the private spaces of the house. The distinct layering of street, forecourt, house and garden at the rear gave a pronounced sense of privacy, differentiating clearly public and private space. It is apparent from the focus of the book that Dennis considers the basis of successful urban design – or the creation of place – must be the kind of spatial matrix that gives priority to streets, squares, courtyards and gardens, as was the case in historic cities.

By the 1970s it had become clear that CIAM's manifesto applied as planning policy was destroying the fabric of much-loved cities. A first significant challenge to the hegemony of modernism came in Aldo Rossi's book *The Architecture of the City*,

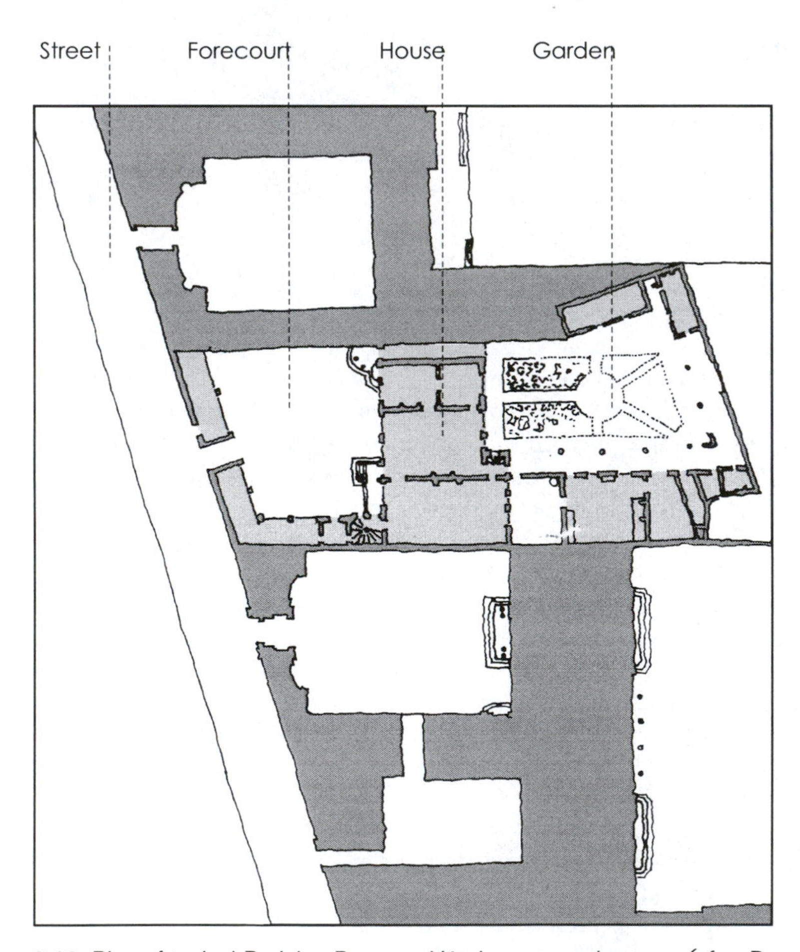

8.13 Plan of typical Parisian Baroque Hôtel – or town-house – (after Dennis).

published in 1966. He rejected CIAM's abstract, functional analysis of the city, proposing instead that it should be seen as a gigantic artefact, something constructed by our fathers and forefathers over time. The city he considered to be the permanent and enduring record of mankind's response to dwelling in the widest sense. It can collect and store memory, can be built upon and modified, and can be read like a historical text.

Major buildings and their associated plazas were, of course, of prime importance to this. But the street pattern of the city itself can contain the memory of its history. For example, the plan of a Roman city can be detected in the centre of Vienna; any cultured Viennese as he or she walks along the Graben towards the medieval cathedral would know that the buildings bounding the left-hand side stand on the foundations of the Roman city walls (Fig. 8.14). The Ringstrasse dotted with monuments marks out not only the nineteenth-century development of Vienna, but also its wide open spaces are reminders of a space kept clear for cannon fire when the city was besieged by the

8.14 Figure-ground plan of Vienna with view of the Graben above. The wall on the left is built on the foundations of the Roman city wall.

Turks in 1529. (The Viennese claim to have invented the croissant, whose crescent shape is a reminder of their vanquishing the invaders.)

Rossi's book led to a revival of interest in Sitte's approach to city planning in the 1970s. Architects tried to make their building footprints adjust to the existing spatial structure rather than impose an object-like form in an undefined space. In *Collage City* Colin Rowe artfully compared Le Corbusier's Unité d'Habitation with the Uffizi Gallery in Florence, showing how traditional architects and planners such as Vasari started by making an urban space rather than a building as a free-standing object (see Fig. 8.15). The Uffizi pushes its accommodation to the edge of its site to make a new street from the Piazza della Signoria to the Arno, whereas the Unité stands isolated in a sea of parking, roads and landscaping.

The renewed interest in traditions of urban design forming the basis of architecture led to the rediscovery of the Nolli plan of Rome. Giovan Battista Nolli went to Rome in 1736, commissioned by Pope Clement XII to prepare a new map of the city[30] (Fig. 8.16). The original feature of his plan was that it showed courtyards and interiors of public buildings, such as churches, as spaces hollowed out of the mass of urban texture. This technique came to be called a 'figure-ground' plan. White emerging from the black background not only prioritised the depiction of space over form but also indicated a layering of space from the most public, civic and monumental to the more private and secretive. Frampton described the primacy given to shaping urban space

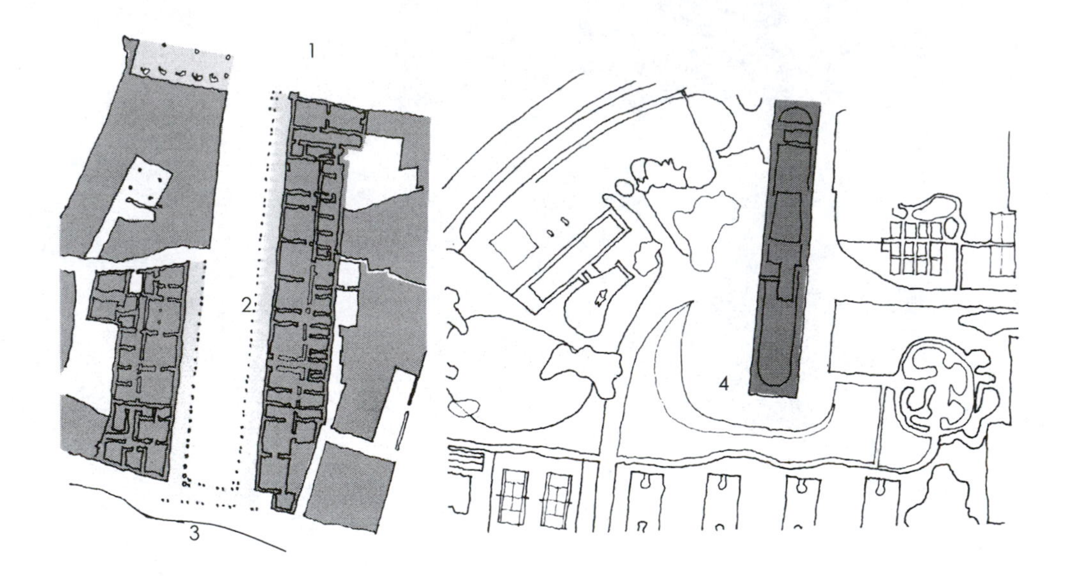

1. Piazza della Signoria, Florence
2. Uffizi Gallery
3. River Arno
4. Unité d'Habitation, Marseille

8.15 The Uffizi Gallery compared with the Unité d'Habitation showing the traditional priority given to space and the Modern Movement's 'object-fixation' (after Rowe).

8.16 The Nolli plan of Rome.

as 'space-form' and is a crucial element of his 'Critical Regionalism' in making specific places that respond primarily to human needs rather than economic determinants. Used in isolation the figure-ground technique can be reductive – as can any two-dimensional drawing – but it remains a good basis for analysing the spatial character of urban environments and is useful as one measure of a design. However, what is of particular interest to us is the extension of a figure-ground plan into three dimensions as this permits a clearer vision of the operation of the environmental forces in urban forms.

Rossi argued that architects should analyse the form of existing buildings and the types of buildings indigenous to their city and design new ones developed from an interpretation of these. Alternatively, a building's form and scale might take their cues from the immediate context of the city, rather than from internal determinates of function. Such an approach lends itself to the repair of our cities and also can be seen as a city-based equivalent of what was discussed in Chapter 2 where looking at vernacular buildings provides lessons for how to respond to conditions of climate and place.

Rebuilding the city

A brief review of three recent schemes where sustainability was a major consideration but where it was combined with clear ideas about urban form will help point the way forward. Coin Street housing on London's South Bank by Howarth Tompkins is a mixed-use scheme comprising 59 dwellings (including 32 houses) and corner shops. The scheme achieves a relatively high density of 334 habitable rooms per hectare (59 per cent above the local planning authority guidelines). A major aim of the architects was to find what they describe as 'an appropriate scale of response'.[31] This entailed striking the right balance between the monumentality required of a city-centre site and the domestic scale appropriate for human dwelling.

A number of different types of layout were considered, including blocks and streets before the architects concluded that a courtyard design best achieved the balance between city scale and dwelling at the same time as providing the potential for a high level of defensible space (Fig. 8.17). Three sides are housing and a community building will complete the court and effectively make a new urban block. The scheme presents an urbane face to the street, its monumentality established by a continuous three-storey perimeter wall of red brick. Some flats have balconies on the street side but these are kept within the bounding wall which reinforces the sense of the urban block being a large single social whole in a way that projecting balconies would not have. Within the block there is a large communal garden designed to cater for a wide range of activities, from protected, overlooked space for children's play to areas for adults to find a quiet retreat from the city hubbub.

Houses with an entrance from the street have a gated, recessed porch raised up two or three steps – this is a threshold detail that reinterprets the Georgian town house described in Chapter 3. The gated porch allows for the front door to be left open on hot summer days for cross ventilation. More typically, sustainable features include timber cladding and other materials to minimise environmental impact, high insulation levels, roof-mounted solar panels to provide hot water and heat-recovery systems.

A second scheme is BedZED (Fig. 8.18a) by Bill Dunster and Bioregional – mentioned in Chapter 4 – which is high-density housing in an outer London suburb based on a reinterpretation of a traditional urban pattern of parallel streets faced with terrace houses.[32] The central residential block of four streets incorporates workshops at the back, the roofs of these providing roof terraces for the dwellings. Bill Dunster has developed this concept of urban planning in the RuralZed project, showing this design in an urban context (Fig. 8.18b). This arrangement produces a street and mews arrangement familiar from Georgian London.

BedZED and its successor aimed for a passive building envelope that reduced the demand for heat and power to the point where it became economically viable to use energy from renewable resources. A three-storey terrace house is in itself more thermally efficient than a typical detached or semi-detached suburban house. All the houses are orientated towards the south with a triple-glazed conservatory or sunspace

8.17 Coin St Housing, by Howarth Tompkins. The top plan shows the courtyard as built and plans below indicating alternative arrangements that were explored.

at the front to provide passive solar heating. Houses incorporate 300 mm super-insulation, floors and walls with high thermal mass, and passive-stack ventilation with heat recovery. A Combined Heat and Power (CHP) plant running off wood-chips supplies energy for the development, which also incorporates power points for a proposed pool of electric cars whose energy is provided by photovoltaic cells. Recycled

8.18 BedZed ((a), photograph copyright Phil Sayer) and RuralZed (b), by Bill Dunster Architects.

steel beams, reclaimed softwood joists and floorboards were a part of a comprehensive strategy for materials that included sourcing from the region wherever possible.

BedZED and Coin Street both show what can be done to make high-density, more compact residential districts with the addition of shops and workshops to leaven the mix. But both were constrained by the limits of what can be done with a single site. The Swedish city of Malmö has recently embarked on transforming a whole district, the redundant Western Harbour area, into a dense, sustainable urban environment known as Bo01.[33] The existing arrangement of streets formed the basis of a new layout of perimeter blocks that were rotated at slight angles and overlapped (such as) to screen and slow the prevailing winds at the same time as introducing a degree of informality (Fig. 8.19). A number of perimeter blocks provide some courtyard housing – not unlike the Coin Street layout – and define a shared open space at the centre of the larger urban block. The new district incorporates two new parks that emphasise the central role that biodiversity plays in the design. Green 'points' that benefit biodiversity are incorporated in the scheme, including bird-nest boxes and bat-boxes and one courtyard is planted with a large variety of native wild flowers.

The aim at Bo01 is to provide all energy from renewable sources: solar collectors, photovoltaics, wind turbines with a large percentage of the heating coming from aquifers and the adjacent sea. In addition to many of the typical sustainable features discussed in earlier chapters, the development has novel aspects such as vacuum

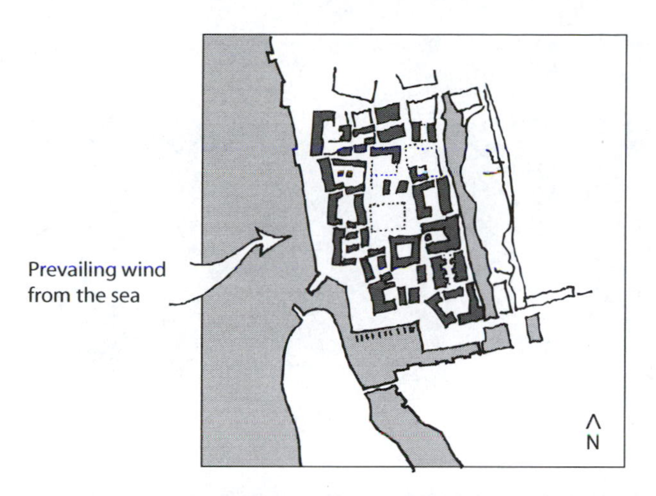

8.19 Site plan of the area known as Bo01 in Malmö, Sweden.

refuse chutes that extract organic waste, some of which is fed to a bio-digester that produces energy and fertiliser. Like Coin Street, the development includes shops and, like BedZED, it also incorporates workspaces to reduce the need to travel.

In their different ways, each of these schemes shows how traditional principles of urban design can assist the development of more compact cities that can readily incorporate, and be enhanced by, sustainable design features normally applied to single buildings. Peter Carl (whose work was noted previously in note 22) argues for a compact city as does Richard Rogers and most – if not all – concerned with urban sustainability. But Carl and Rogers appear to have opposing views as to what its character should be. Rogers is optimistic about the capacity of rapid technological advances to deal with the environmental problems we face and the changes we must make. 'We must pursue even more decisively the development of technologies,' he says, in pursuit of his vision of the future.[34] For Rogers, two of the main problems are an unbridled market economy and a reluctance to wholeheartedly embrace change.

Carl, writing from a more historical point of view, states that the 'city is a permanent receptacle of metamorphosis'. He points out that there is nevertheless 'always an element of recollection, a dialogue between what a city has always been and the most recent transformations'.[35] His research identified the stabilising role of the urban block as the key to allowing change whilst fixing memory. (The influence of Rossi's thinking is evident here.) Although urban blocks vary in size and scale from city to city, Carl identifies a common characteristic of 'several layers of depth' from street to the interior of a block.[36] An urban block might contain courtyards of various sizes and a range of activities from private dwelling, gardens, shops, offices, even a university department in the block he studied in Padua.

Analysing the urban morphology of any given city to understand how it has evolved to respond to both changing aspects of living and permanent or enduring ones, would seem to provide one starting point for designing the sustainable city. A good example is the way a long concrete steel and glass canopied colonnade is used in the new Scottish Parliament building to 'tidy up' a fragmented urban block. In addition to making a grand entrance, it also defines a new urban space of civic importance at one end of Edinburgh's Royal Mile. The other end of the long colonnade opens into

8.20 The Scottish Parliament, Edinburgh, by Enric Miralles (EMBT/RMJM).

Arthur's Seat, thus dramatising the experienced relationship between city and nature (Fig. 8.20). Adding photovoltaics to the suitably orientated parts of such an extensive canopy would make this a useful sustainable feature as well as an engaging urban design element that would open the periphery of the building to its immediate context and to passers-by. Accepting the urban block as a basis, the technological advances advocated by Rogers could be accommodated – as well as the many points made here in earlier chapters – and focused on obtaining more daylight, natural ventilation and saving or generating energy for what some ecologists insist will have to be the compact cities of the future.

Keeping in mind the idea of urban design as making the setting for our lives, we will conclude this chapter with an environmental overview at the city scale.

An environmental overview

The journey from the Acheulian hut of Figure 2.2 to the modern city of Figure 8.21 may have taken 350,000 years, but the laws of physics haven't changed during that time. Radiation, convection and conduction are as important as ever and similar phenomena occur in both camels and cities. What modern urban conurbations do, however, is take in extraordinary (and unsustainable) quantities of fossil fuels, materials, food and water and produce enormous amounts of waste materials, heat and light. In doing so they become significant modifiers of climate – their own temperatures can be 5–10°C higher than their regions and the pollution due to CO_2 production in cities is a major cause of global warming. Nonetheless, it must be emphasised that the city is a place of potential – for example, the annual solar energy falling on London is about ten times the energy demand of the city.[37]

The environmental relationship between city and region has tended to be one of master and servant with the servant coming out rather poorly, in the customary way. Mozart's Leporello declares that he 'no longer wishes to serve'; however, even the death of his master Don Giovanni doesn't quite free him from his bonds. The regions around major cities with their unattractive (and potentially dangerous) infrastructure,[38] extensive landfill sites, incinerators and polluted waters also have not yet gained their freedom. The exceptional energy demands of cities are clearly seen in the scale and extent of the pylons and overhead cables that march from outlying power stations to cities. This past scarring of the landscape for some finds its current counterpart in

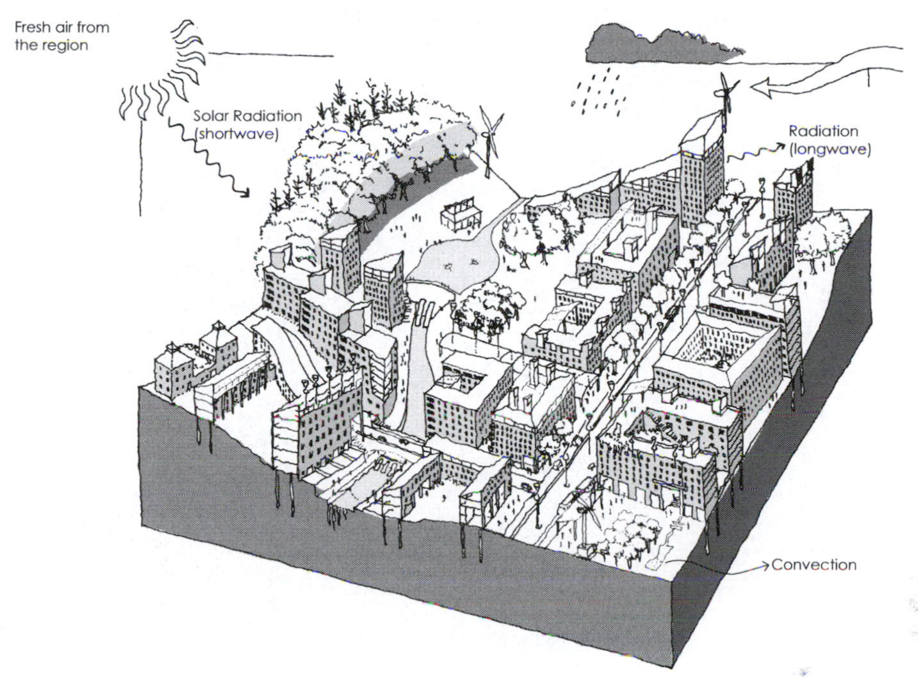

8.21 Energy in the city.

proposals in France and England to create wind farms in rural areas to serve cities. However, there are wind turbines and wind turbines and each site is different. Hills and moorlands may well be inappropriate but other areas such as the disarticulated peripheries of our cities might be suitable for large turbines. Although not as efficient as on exposed moor or hilltops, this would be partly offset by a reduction in transmission losses. The scale of buildings and industrial infrastructure on twentieth-century urban edges would accommodate such turbines. These areas are shaped by fast road systems and are focused on shopping, work or sport. These spaces of transit and distracted interiors contrast with the quiet contemplation that people find when in the countryside. Imaginatively used, wind turbines could even be used to 'urban design' these fragmented and characterless spaces, bringing order as does a line of dockyard cranes (Fig. 4.24).

When mechanisation gained command it took control not only of our buildings but, of course, of our cities too. This can be clearly seen in form 'A' in Figure 8.22 which shows the energy demand in Sydney in 1990–91 as a function of temperature.

At lower temperatures more energy is required for heating and, at higher ones, more for air-conditioning. What all of this does, of course, is add heat to the area and cities themselves, thus modifying the local climate (and that of the world).[39] What we need to do is to lower the energy consumption at all temperatures[40] and create a much broader 'valley' as shown in form 'B', thus, weakening the control that (fossil-fuelled) mechanisation exercises. And then we need to make better use of energy, materials, water and 'wastes'.

The first issues are those of scale. It is important to have an idea of the respective roles of city and region. Our view of this very political question is that there should be

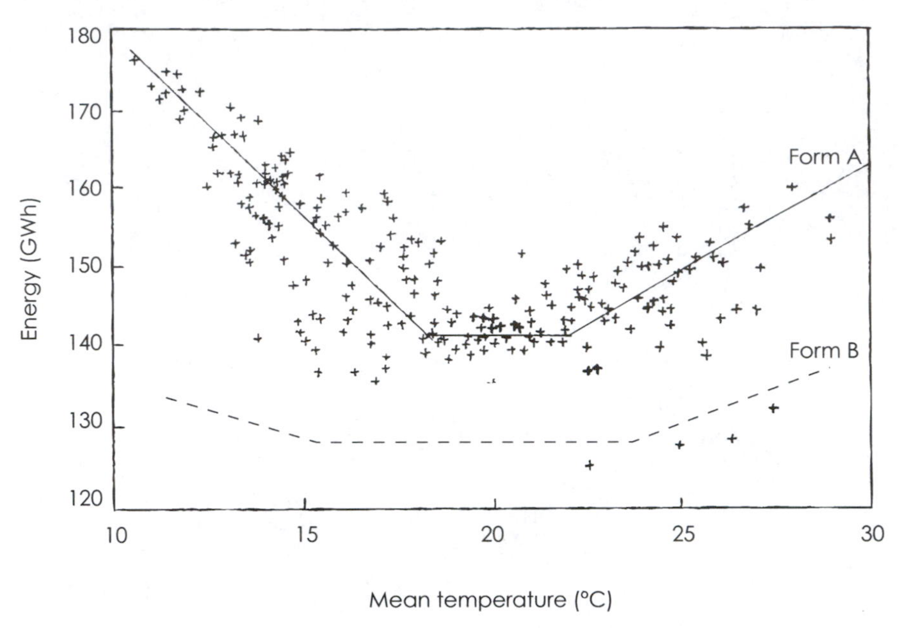

8.22 Energy demand and temperature in Sydney, Australia.[41]

a 'good-neighbour' policy and that cities should achieve as much as they can within their own boundaries. Of course, geology and ecology (and economics) ignore the city's demarcation lines and policies, such as those for water and biodiversity, must be based on a broad view – one that you might see, for example, from the window of an airplane. Some environmental issues are best treated at regional level, others at that of the city, district, neighbourhood or block. We may find that in the future the regions grow crops to supply cities with both food and energy.[42] Densities in cities can offer important potential economies of scale and the scope for reduced energy use for transport and the potential for efficient, compact energy and material networks are considerable. If we think of cities as complex and compact organisms, their physiological systems require less material to serve a given volume. Often, we ignore (or forget) how cities function, but when the services are run above ground as in the case of the main district heating pipes[43] shown in Figure 8.23 in Astana, the capital city of Kazakhstan, it is difficult not to pay attention to a key element of the city's 'metabolism'.

At what scale heating or power or water treatment or any other service should be provided is a moot point and will vary with circumstances ranging from local costs to available technologies.

However, unless we get the city scale right we won't be able to make sufficient progress in sustainability. The biological analogy might be that any one building (or perhaps category of building such as housing) is just one organ in the body. Survival depends on all organs working together.

The urban challenge is to reduce environmental impact and meet the essential demands in humane, fair and sustainable ways. This has major implications for servicing, planning, morphology and design.

8.23 Heating pipes in Astana, Kazakhstan.

Servicing

For their servicing, cities for the most part now function in linear, rather mechanistic, ways (Fig. 8.24). Inputs of goods, water, energy and food are transformed in the least imaginative ways possible into a series of waste products.

Natural systems, on the other hand, are integrated wholes that use materials and energy cyclically and efficiently, thus minimising waste. Plants and animals die – micro-organisms help decompose the wastes and are rewarded by being eaten by other organisms as the next step in maintaining a cycle that has been functioning for many hundreds of millions of years.

A more 'ecological' city will take the same inputs (but in smaller quantities) and recycle more of the 'waste' products in the city and the immediate region (as in the case of the nineteenth-century Parisian suburbs mentioned before). It will also provide its own energy in a variety of ways, using its buildings, roads and green spaces and by recovering energy from human and plant wastes and materials. We can expect that cities (with their regions) that are able to do more with less and are capable of change will be able to adapt better to a world under environmental pressure just as organisms adapt to survive.

One way of considering the impact of cities is the concept of an ecological footprint.[44] In Chapter 2 we discussed a building's physical footprint – the shape and size of mark that it makes on this site. The ecological footprint is different in that it is meant to be a measure of the environmental impact. It has been estimated that the current

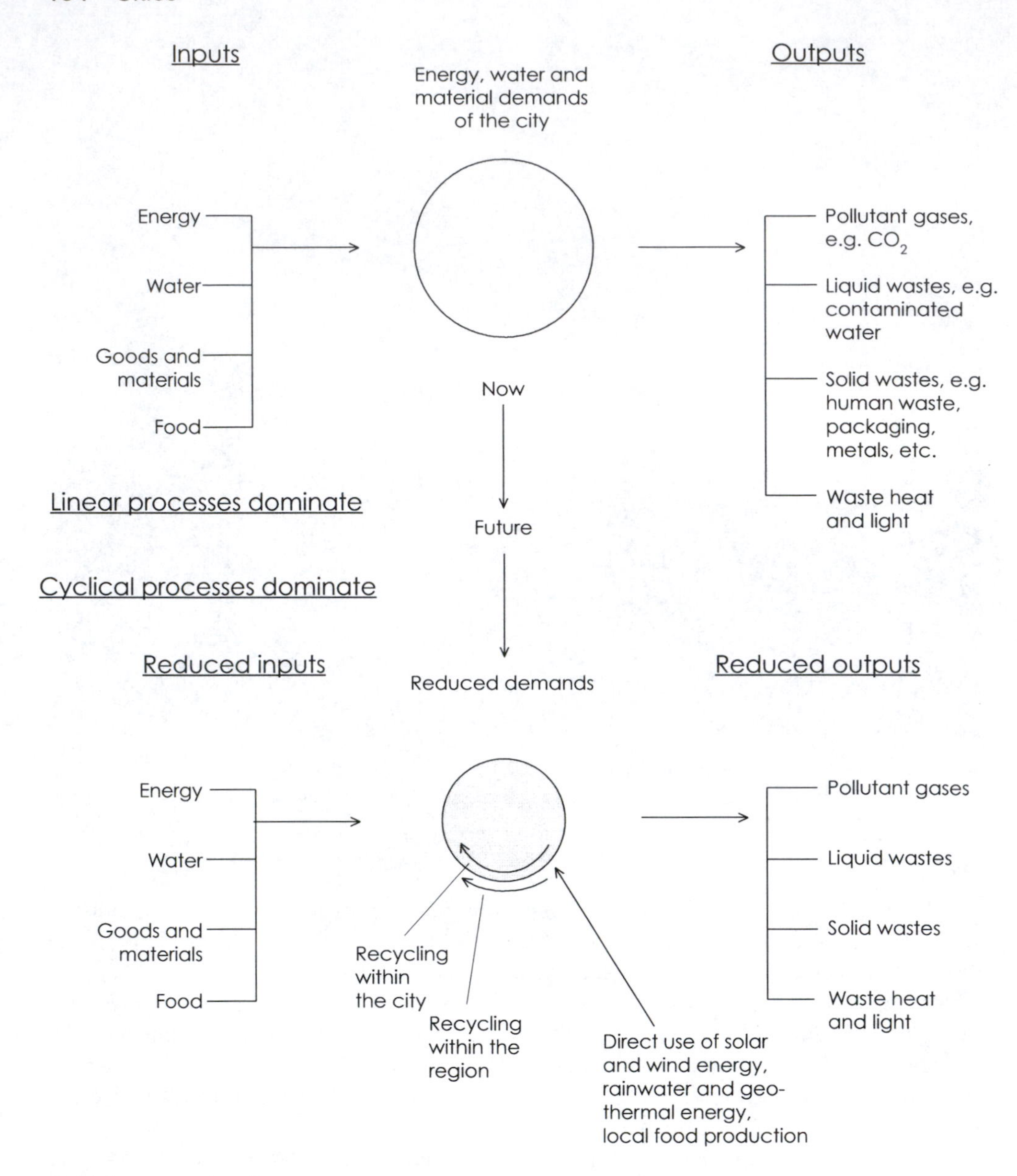

8.24 Linear and cyclic processes.

way of living in the UK, if extended to the rest of the world, would require three planets rather than the one currently available and it has been calculated that Londoners have an ecological footprint about 5 per cent higher than the UK average.[45] Needless to say, this is not sustainable.

The shift towards a more 'biological' city with less of an environmental impact will involve viewing it differently, and carefully examining all of its potential from the ground to the air. Urban areas are three-dimensional and need to be considered as such. A very simplified way of seeing the city is as a complex organism operating in three verti-

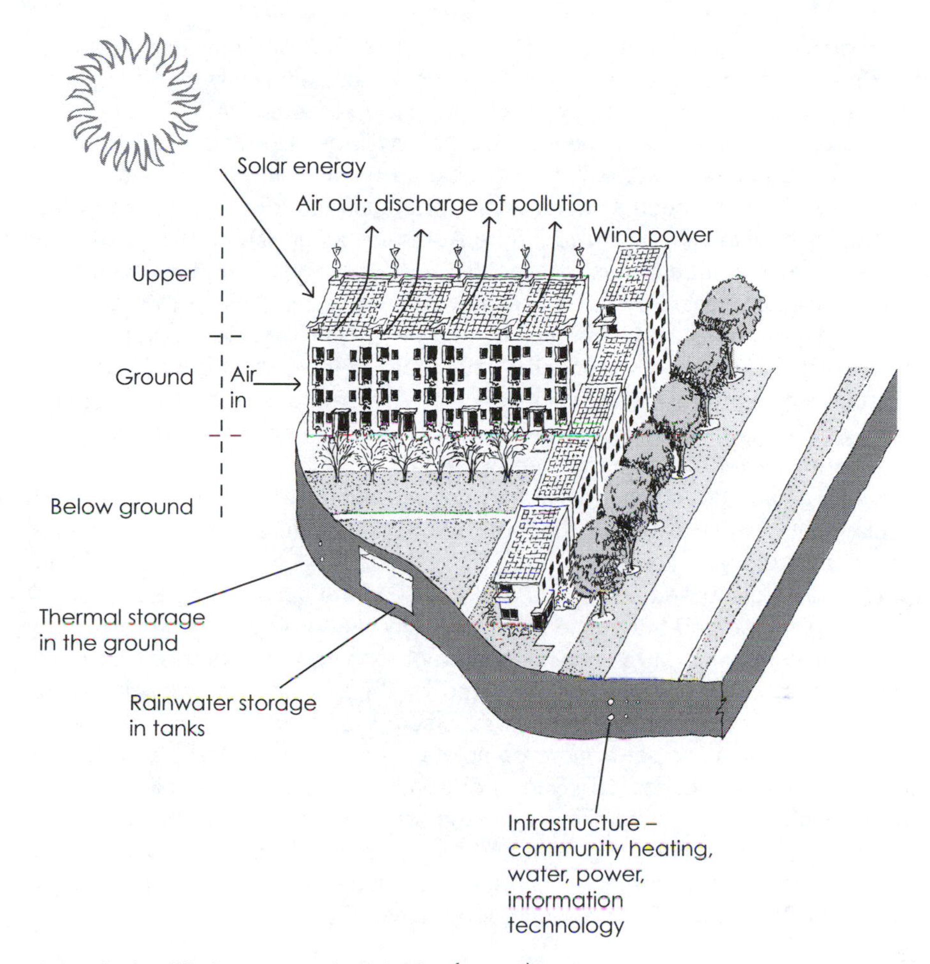

8.25 A simplified conceptual model for (part of) a city.

cal zones with an important horizontal spatial organisation. Figure 8.25 shows a simple model of this.

In the upper zone, solar energy and wind energy will be captured and rainwater will be 'collected'. Supply air may come from here or the ground zone but stale air is likely to be exhausted in this region. This upper part is a bit like a forest canopy.

In the ground zone, buildings will be serviced by, among other sources, solar energy and rainwater. Much of the distribution will take place here. This will be done as efficiently as possible. These buildings are a bit like stationary, homeothermic camels. Green spaces will play many roles, including providing opportunities for social contact, leisure and tranquillity. They can be both a refuge from harsh environmental conditions (as were early Persian gardens which provided relief from a blistering sun) and beneficial modifiers of the local climate. The latter will include reducing noise levels, filtering the air and producing oxygen and helping to lower the urban heat build-up and so reduce the need for air-conditioning.

Below ground will serve as a zone for distribution (large-scale infrastructure) and for storage of collected solar energy and rainwater. The infrastructure is likely to include more efficient community heating schemes which include combined heat and power (CHP) which, as noted before, can use energy more efficiently than the separate power and heating systems largely used now. (The pipes shown in Figure 8.23 are part of a community heating scheme – at the city level. BedZed has a smaller scheme, as does Coopers Road, which is discussed below.) In the longer term we can anticipate a shift to hydrogen as a major part of the infrastructure. Because of the temperature differential in this zone (see Figure A.4 in Appendix A), it will be possible to use the energy as a source of heat or 'coolth' (see Glossary); using the roads as a heat source is another possibility.[46] The lower volume is a bit like the root zone of plants but also has some of the characteristics of animal burrows. Georgian planning in eighteenth-century Bath used this underground zone. At the Circus, John Wood the Elder had an underground reservoir which was higher than the level of the basement kitchens and which provided water to the houses by gravity feed. Two-hundred-year-old plane trees now cover the reservoir (see Fig. 8.26).

In the future we can hope to see (or, rather, not see but have) large-scale insulated stores of solar-heated water serving a modern (urban) architecture of similar outstanding integration and quality.

The vertical zones discussed above find a response in the house. Alberti described the city as a large house and the house as a small city. According to Bachelard[47] the house can be imagined as a psychological construct ranging from cellar to garret (or attic) as we have seen. Dark fears are prompted by thoughts of the cellar, he says, whereas poets traditionally inhabited attics, free to soar like birds.

As well as this vertical zoning, a more complete way of living in the city might be furthered by an equivalent horizontal layering of space. A distinct threshold between the public realm and the private domain – as in the steps up over the sunken area of a Georgian house, or the small front garden and recessed porch of a Victorian terrace house – can respond to the psychological balance needed in our lives. The deeper into the dwelling the more private and secure it becomes.

8.26 The Circus, Bath.

Planning

In *Through the Looking-Glass*, Alice repeatedly sees one sign labelled 'To Tweedledum's House' and another 'To the House of Tweedledee'. The sets of signs point in the same direction and she surmises (correctly) that they live in the same house and that she is on the path to one place. Our path is similarly clear, and we are moving towards solar cities based on energy from the sun, wind and biological materials.

In these cities rigid, mechanistic concepts of zoning will be replaced by a greater integration of all urban activities – work, housing, recreation and transport will overlap. The monolithic blocks of the Modern Movement will give way to compact, multi-use neighbourhoods. The continuity between the natural and built worlds will be strengthened;[48] green spaces will play many roles as noted above. Landscaping and vegetation will provide pleasure,[49] a distinct sense of locality and an aesthetic structure for the city.

Movement networks will favour pedestrians, cyclists and mass transport and if we are fortunate the city will be wrestled away from the car so that it can develop in richer, more diverse ways. The creation of such a sustainable transport infrastructure should result in less area required for roads, more opportunities for a more humane public realm, less energy use and reduced CO_2 production.

Morphology

Geology, topography and the need for movement, light and air shape the city. Its height rises with advances in materials science and technology (both for the structural and 'physiological' systems). And, of course, as Rossi and many others have noted, urban structure and form are the products of social and cultural processes at every level. (Rossi's work for example, draws on his intimate knowledge of northern Italian cities and their typical building forms.)

The sustainable city of the future is likely to be more compact than now, although a balance will need to be struck between density and amenity. Even in eighteenth-century Paris the concentration of activity there meant that the city was losing its human scale – overcrowding was growing and movement could be dangerous. Collisions between coaches and pedestrians were so common that a standard scale of fines for injuries to legs, thighs and arms was codified.[50]

Such compact forms also have less surface area per unit of volume enclosed, which means less area for solar collection. This is an important consideration as we move towards solar cities. (Many of the morphological aspects of cities are similar to those previously discussed for buildings and animals. Some phenomena are related primarily to surface area, for example, heat loss and the collection of solar energy and rain water; others particularly to volume, for example, energy demand; and, inevitably, in some cases both play a part.)

There is a need to orientate the buildings, and arrange them to capture the sun's energy for heating, daylighting and production of electricity (as discussed previously).

Just as nineteenth-century Paris successfully developed a morphology which guaranteed rights-to-light by controlling street-width separation between buildings and building height (see Fig. 6.15), similar rights to solar energy will become widespread.

Before in Figure 5.15 we saw an elegant example of an urban housing scheme in Belfast designed to make use of the sun's energy with particular attention given to using photovoltaics on the roofs and to presenting an elevation of interest to traffic (and pedestrians) on the main road to the east. (The project included a nine-storey tower block and shows that it is possible to combine high-rise buildings and environmental sustainability if careful attention is given to orientation, form and layout.) Maximising the solar potential is best carried out through detailed modelling. Figure 8.27 shows an illustration of how the solar radiation incident over the course of a year on complex urban forms in London can be determined. Modelling can also be carried out to assess the potential for the use of wind energy in relation to various urban morphologies.

The results of such functional studies should be integrated into designs without dominating them. We can probably have a great variety of forms and patterns which allow us to use renewable energy sources. And with time we can expect to see cities adapting. They will become more like organisms interacting with their environments by adjusting their street patterns, building groupings and forms to derive more from the resources of their sites.

Similarly, their 'physiological' systems will change and their circulation, energy and material networks will adapt.[51]

Just as we saw in Chapter 7, morphology will in part be a reaction to local light and temperature conditions. This combined with attention to wind should result in regional variations of the form of our cities.

Design

Environmental considerations will undoubtedly change current views on the design of cities. The art and science of creating places which are varied, humane and sustainable is in its infancy (although humane and beautiful cities form part of our legacy, of course). Perhaps it starts with geology, moves on to peoples, culture and history and only then addresses architecture. It would probably benefit from blending the kind of site analysis discussed in Chapter 2 (topography, temperature, wind, incident radia-

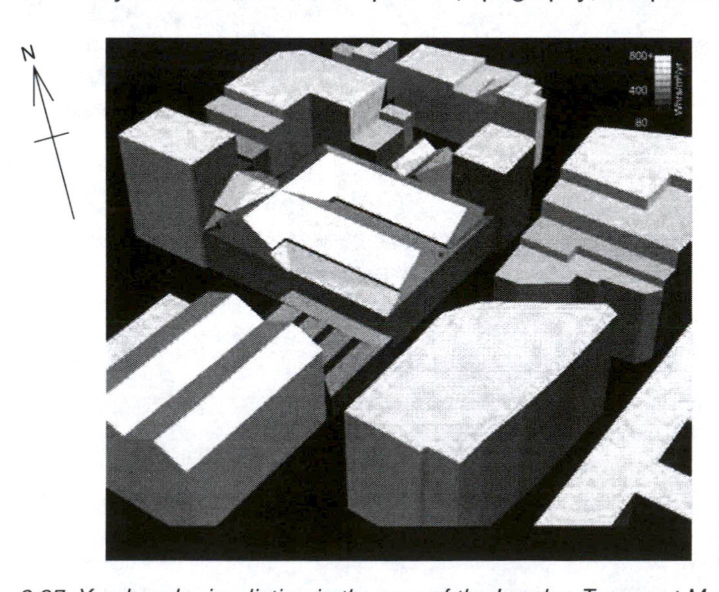

8.27 Yearly solar irradiation in the area of the London Transport Museum.

tion and so forth) with more traditional approaches, such as discussed above, which look, for example, at street patterns, scale and massing and landmarks.

Building design is intimately related to the locality, environment and setting of the city. To give but one example, in urban areas that are less polluted and noisy it is easier to use natural ventilation. (In London where penalties have been introduced to discourage traffic in central zones, people have been able to open office windows for the first time in years.)

Architectural style, however, is likely to be a secondary issue compared with creating an urban fabric which has both physical and biological components. One way of viewing parts of the twentieth-century city is as a direct transformation of traditional urban areas – buildings became space and space became buildings, and to maintain the density the buildings had to go higher. This can be seen in a modified three-dimensional version of the Nolli plan discussed previously (Fig. 8.28).

Both of these approaches can probably be made reasonably environmentally sustainable. However, as noted above the 'feel' of the urban experience is very different – the importance of the street, the perception of the extent to which the buildings are firmly rooted in the urban fabric, the opportunity to appreciate architectural detail, the degree of social contact are just a few examples of what varies. To us the tower block seem less sociable and humane.

Generally, we can expect to welcome more variety in design. Forms may be more articulated for sun, light, ventilation and shelter from noise. Façades will be more responsive to their environment. Roofs will be planted with vegetation (for both environmental and aesthetic benefits) or painted white (to reduce heat gain) or offset from the building grid to provide an area for solar collectors (as suggested by the non-solar expressionist Munich roof of Figure 8.29).

Amidst all of this design, it should be possible for unplanned beauty to exist.

Manchester and London

For an important competition for the design of a sustainable community in Angel Fields, Manchester, one team including one of the present authors[52] set out to investigate the potential for combining an environmental approach with good architecture.

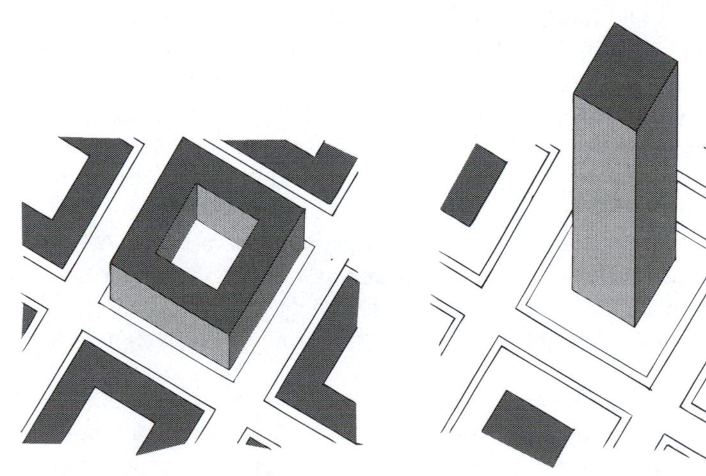

8.28 The city transformed.

8.29 Munich roof.

There was a shared belief that city dwelling is enriched if brought back in touch with nature and, ultimately, that the next age of design will be more biological than mechanical. Instead of a house being a machine for living in (*machine à habiter*), the city could become a combination of urbane spaces and park-like spaces with, perhaps, more parks than at present.

The team's goals were a lively, varied, sociable space with a positive relationship to its physical and biological environment, and a place where they would like to live. The first manifestation of this was a series of undulating green strips (Fig. 8.30a) representing the roofs of buildings running north–south. These played a symbolic role, arguing for a more organic, softer approach to urban living than has been common before. The view was that environmental considerations should inform the urban morphology, height, massing and so forth, but not determine it uniquely.

Courtyard strips running east–west and north–south were compared. Although detailed calculations were not carried out it seemed that as long as the key criteria of being able to use the roofs for photovoltaics (PVs) and active solar collection, and spacing the buildings for ample daylight were met, both forms were potentially suitable.

This approach argued for putting taller buildings to the north (Fig. 8.30b) and avoiding overshadowing. By this stage the site was being seen as a tapestry of warp and weft, with additional tension added by interrupting the pattern to create discordant notes. The strands of the tapestry included buildings, water flowing across the

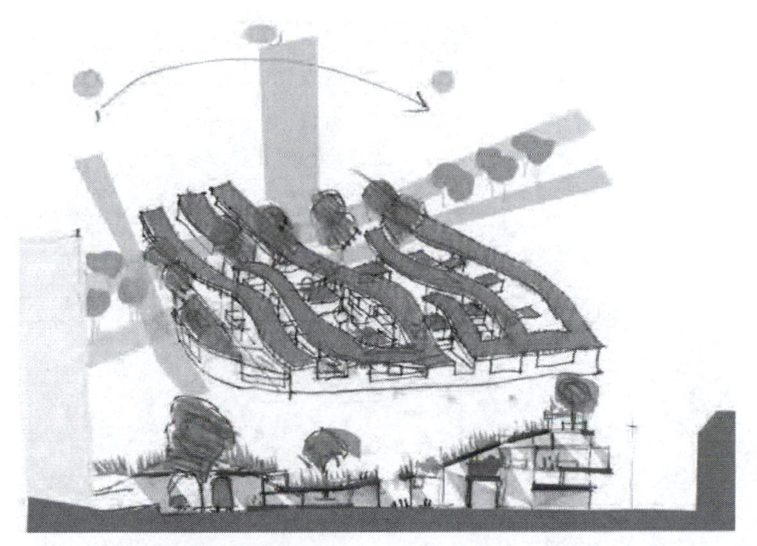

a Concept sketch.

b Developed scheme.

8.30 Angel Fields.

site, small and large gardens, pedestrian paths and bicycle routes. (Needless to say, all of this is not evident in the illustrations which were but a start.)

By designing for extremely low demand, by providing a place where energy, water and materials can be dealt with (including the cyclic process of a methane digester for organic waste from the site and neighbourhood providing gas) and by planning for elegant vertical wind turbines, it was estimated that more than half of the building energy demand could be met on site.

Coopers Road Housing, London

A more traditional urban pattern is found in the perimeter block structure with inner courtyards at the Coopers Road housing scheme by ECD Architects (Fig. 8.31).

After careful consideration it was decided to demolish a 1960s' tower block complex and replace it with the forms shown. The decision was partly based on the

8.31 Coopers Road Housing, London.

difficulty of upgrading the 1960s' scheme to the much higher environmental standards of today.

What is striking is that this urban renewal scheme is a reversal of the transformation of the city that occurred in the mid-twentieth century (as shown notionally in Figure 8.28). One advantage of the more sociable courtyard structure of Coopers Road is that it is easier for parents to keep an eye on their children in the playgrounds. The form is well suited for good daylighting and natural ventilation. The environmental strategy includes community heating with combined heat and power (CHP) and provisions for the use of solar energy and rainwater recycling.

Conclusion

What the case studies cited before, the Manchester competition entry, and Coopers Road point towards is another conception of the city based both on good design and biological and ecological principles.

What might our cities be? Philosophically, let us take an Aristotelian view that they should make us happy and secure. For us to be happy, they need to be lively, vibrant, diverse, interesting, beautiful, peaceful and one thousand other things. This will require an architecture and urban design with people in mind. As suggested before it will be Janus-like, looking both backwards to history and forward to the opportunities that progress brings. (And it will need to look with a critical eye in both directions.)

For security we want to not only be able to walk the streets safely but to know that our cities are a benefit for our environment rather than the cause of its destruction. A world menaced by global warming, uncertain energy supplies, insufficient water and the other modern horsemen of the Apocalypse is an insecure one.

Aristotle's criteria can be satisfied – and in some respects the cities that do so will not be that different from the ones he knew. The principal question though recalls Marvell's line 'had we but world enough and time ...'. World enough we probably have but as to time we shall see.

Further reading

Kostof, S. (2001) *The City Shaped*, London: Thames and Hudson.
Pearce, F. (2006) 'Ecopolis Now', *New Scientist*, 17 June, pp. 36–45.

Building references

Coin Street: Architects – Howarth Tompkins; Environmental Engineers – Atelier Ten; Structural Engineers – Price and Myers.

BedZED: Architects – Bill Dunster Architects; Environmental Engineers – Ove Arup and Partners; Structural Engineers – Ellis and Moore.

The Scottish Parliament: Architects – EMBT/RMJM; Environmental Engineers – RMJM Scotland/Buro Happold; Structural Engineers – Ove Arup and Partners.

Coopers Road Housing: Architects – ECD; Structural Engineers – Price and Myers; Environmental Engineers – Max Fordham LLP.

9

Conclusion

We have tried to show how functional considerations influence the design of both biological forms and buildings and how context, history, people and time contribute to the creation of our architecture and cities.

This book is part description and part plea and we have set out a few ideas of what we believe to be important. Functionalism alone is inadequate for the creation of a meaningful architecture. Proust refers to the 'enormous edifice of memory' and we share his view of its significance. As we have said, architecture and urban design need to be located in an environmental, historical and cultural context. The environmental crisis creates an opportunity for a varied architecture which responds to site, setting and context. Our approach offers, we hope, the possibility of a humane way of living that is in harmony with the planet.

A combination of a general or biological principle, an example either from biology or from vernacular architecture and a contemporary application, has been used at times to illustrate our views. In living organisms and vernacular buildings we see a variety of ways of successfully interacting with the environment. The vernacular offers us lessons in how to respond to context drawn from many centuries of observation and practical adjustments – the result is often buildings which create a sense of place and reflect their culture. They speak of the importance of the human being and the community.

A number of major themes are woven into the text. One is unity – there is often a unity of site, setting and context which can be distinguished at a variety of levels in the cultural patterns of an area. Then, too, the laws of physics create a unity from site to city and as we have seen the same processes apply to prehistoric huts, camels and urban configurations. What is important in a house is important in a city. The physical and psychological comfort we find in our immediate surroundings should be found in our larger 'homes' – our neighbourhoods and cities.

We have traced a period of some 350,000 years. During this time the population (*Homo erectus* at first, *Homo sapiens* now) has gone from under 100,000 to 6.5 billion. By mid-century as mentioned before, we may be nine billion with more than two-thirds living in cities. Individual buildings in the city need to focus on human dwelling. The city itself should be primarily considered as the environment in which we (and a whole variety of other species) live rather than a set of abstract processes of organisation or form.

Evolution and adaptation have influenced our approach. Just as nature evolves by using the same materials (we share 30 per cent of our genes with daffodils and it isn't that long ago that we were fish), our buildings and cities will evolve by a basic reliance on similar materials and forms. And these structures, just like organisms, will need to adapt to survive. Part of this process will be the shift from a period when mechanisation was in command, a time when technological development provided what felt to some to be like unlimited power, to a more holistic biological approach. Technology alone is inadequate to provide us with the basis for a sustainable architecture and we have no unerring faith in it. Hiroshima remains in our memory – Albert Camus said of the atomic bomb that it marked the highest degree of savagery of the mechanical age. Chernobyl is all too recent.

What form will these buildings take? A wide variety. Vernacular buildings were often passive and defensive in the way they modified environmental conditions. Modern technologies and materials allow buildings to become more open to their surroundings – more 'active'. We have seen that energy and particularly the capture of solar energy can influence form. However, the effect can be of varying importance. It is likely that one branch of environmental architecture will be a very distinctive one in which functional characteristics are strongly expressed, for example, the Contact Theatre. (This can be linked, of course, to a historic tradition as it is with the reference to Viollet-le-Duc's rationalisation of Gothic architectural principles to embrace the possibilities presented by iron and steel in buildings.)

Our future, we believe, depends on renewable energy sources and as these vary with locality, there is a clear basis for a 'critical regionalism' with an architecture which is diverse and distinctive.

We hope that this book will encourage a transition to a more biological, environmental and historically informed architecture which is rich in meaning and which will provide us with the diversity we crave. Proust said that 'the future is what we project of the present'. Our view is that it is what we project of both the past and the present. May it be better than both.

Appendix A

Earth, sky and physics

> ... to see the world in a grain of sand.
> (William Blake)

One or two reasons why life on Earth is possible

On the Moon, which has no atmosphere, the temperature rises to 100°C on the sunlit surface and drops at night to −150°C. The atmosphere of the Earth is like an insulating blanket that smoothes the diurnal variation and results in an average surface temperature of about 15°C, creating conditions much more suitable for us. Figure A.1 shows energy flows for the Earth schematically, similar to others we have seen earlier, for example our ever-popular camel of Figure 5.1.

Very simply, part of the incoming solar radiation reaches the Earth's surface which in turn acts as a radiator. However, the infrared radiation emitted by the Earth can not freely escape into space because it is absorbed by water vapour and carbon dioxide (and a few other gases) in the atmosphere. This heat warms the atmosphere and in

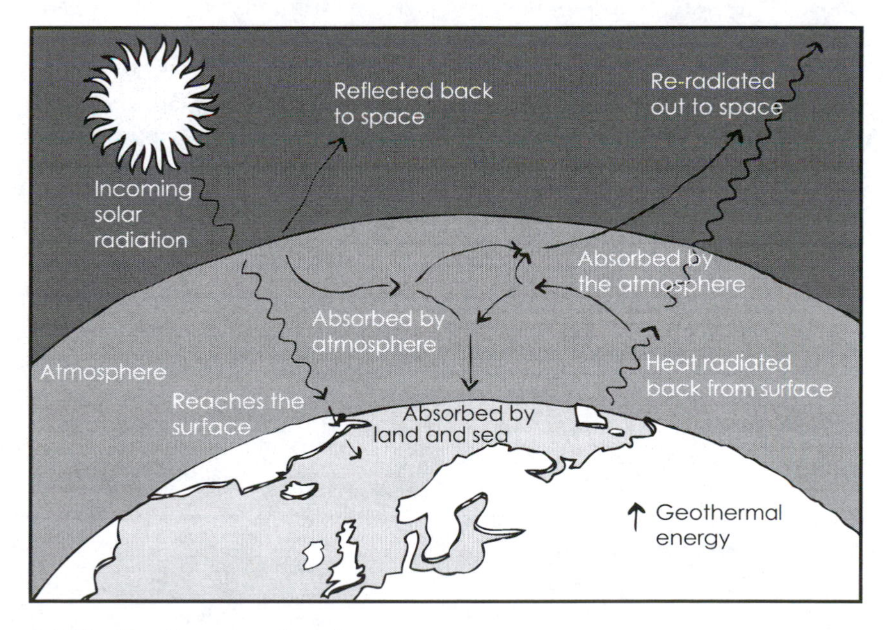

A.1 The Earth's energy flows.[1]

turn it radiates heat back to the ground keeping it warmer than it otherwise would be and this is referred to as the 'greenhouse effect'.

Over the past century and a half or so the burning of fossil fuels has led to increased CO_2 levels and so an increase in heating of the Earth's surface i.e. global warming. The dangers that this presents are known to almost all, an apparent notable exception being George Bush Jr in the first years of the twenty-first century. It has been calculated that what is happening is the equivalent of about an additional 1 W/m^2 everywhere on the planet.[2] To give this some context, if a human body were subject to the same energy imbalance it would reach a non-sustainable temperature of 50°C in under two weeks.

For the Earth it is predicted that this will result in a temperature rise of about 0.3°C in the next 30 to 40 years.[3] Such a temperature rise is similarly unsustainable. The possible consequences include melting glaciers in the Himalayas, melting of the polar ice sheets, rising ocean levels, saline contamination of coastal freshwater supplies – we could go on and this reminds us of the fragility of our world.

To return, another reason why life has been possible is that from approximately 2,500,000,000 to 400,000,000 years ago the level of oxygen increased due to the activity of cyanobacteria permitting the formation of an ozone screen in the upper atmosphere. The ozone layer protected organisms from damaging UV radiation and allowed the emergence of life on land, perhaps in the form of algae[4] – the rest is history. (Recently, the ozone layer has been damaged by CFCs and other chemicals. Legislation is currently in place to attempt to limit further destruction. It is worthwhile keeping in mind the architect Tadao Ando's phrase 'the Earth is crying' and in the future do better.)

In early winter it is sometimes beautiful

The thermal mass of the water has a moderating effect on extremes of land temperature.[5] This is suggested in Figure A.2 which shows coastal and ocean temperatures on the northeast coast of the USA.

The effect, of course, is similar elsewhere. On 17 October 2003 Britain experienced glorious sunshine with mid-day temperatures of about 16°C in the south and on such days one is tempted to ask as Shakespeare did in *Cymbeline*, 'Hath Britain all the sun that shines?' (Unfortunately, the answer usually is only too obvious.)

The reasons for this particularly sunny day are related to geography and the position of the Isles near the Gulf Stream and at a more fundamental level to physical principles of thermal mass, moisture movement, wind and solar radiation – all considerations that we have seen are integral to building and urban design. Wind, after all, is merely air in motion.

The situation is summarised in Figure A.3.

A large high-pressure system (1032 mbar or 103,200 Pa) sitting to the south of Scandinavia brought in dry air and blocked out wet depressions (i.e. low-pressure areas) from the Atlantic. Around this high-pressure system were chilly easterly winds coming from the continental land mass which in October cools down rapidly as it loses the heat stored during the summer.

When the great American engineer and inventor, Willis Carrier, developed air conditioning he had the insight to call it artificial weather.

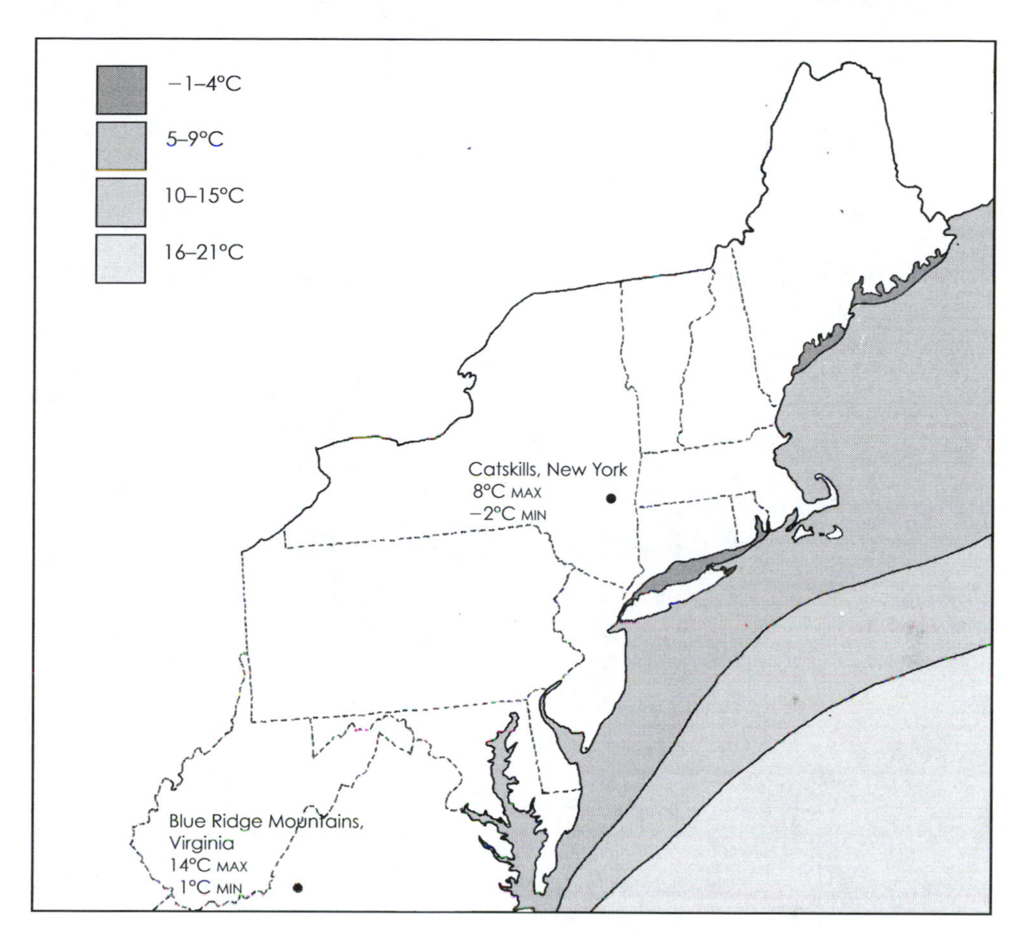

A.2 Land and coastal temperatures on 29 January 2006.[6]

Temperature

Diurnal range

The diurnal range is important in sustainable architecture because it gives an idea of the potential for night cooling and hence natural air conditioning systems. Table A.1 gives data for a few cities.

Table A.1 Diurnal range in selected cities (July 2003)[7]

	Average daily low (°C)	*Average daily high (°C)*	*Diurnal range (°C)*
Paris	14	24	10
Hong Kong	25	30	5
Singapore	24	31	7
Cairo	21	35	14

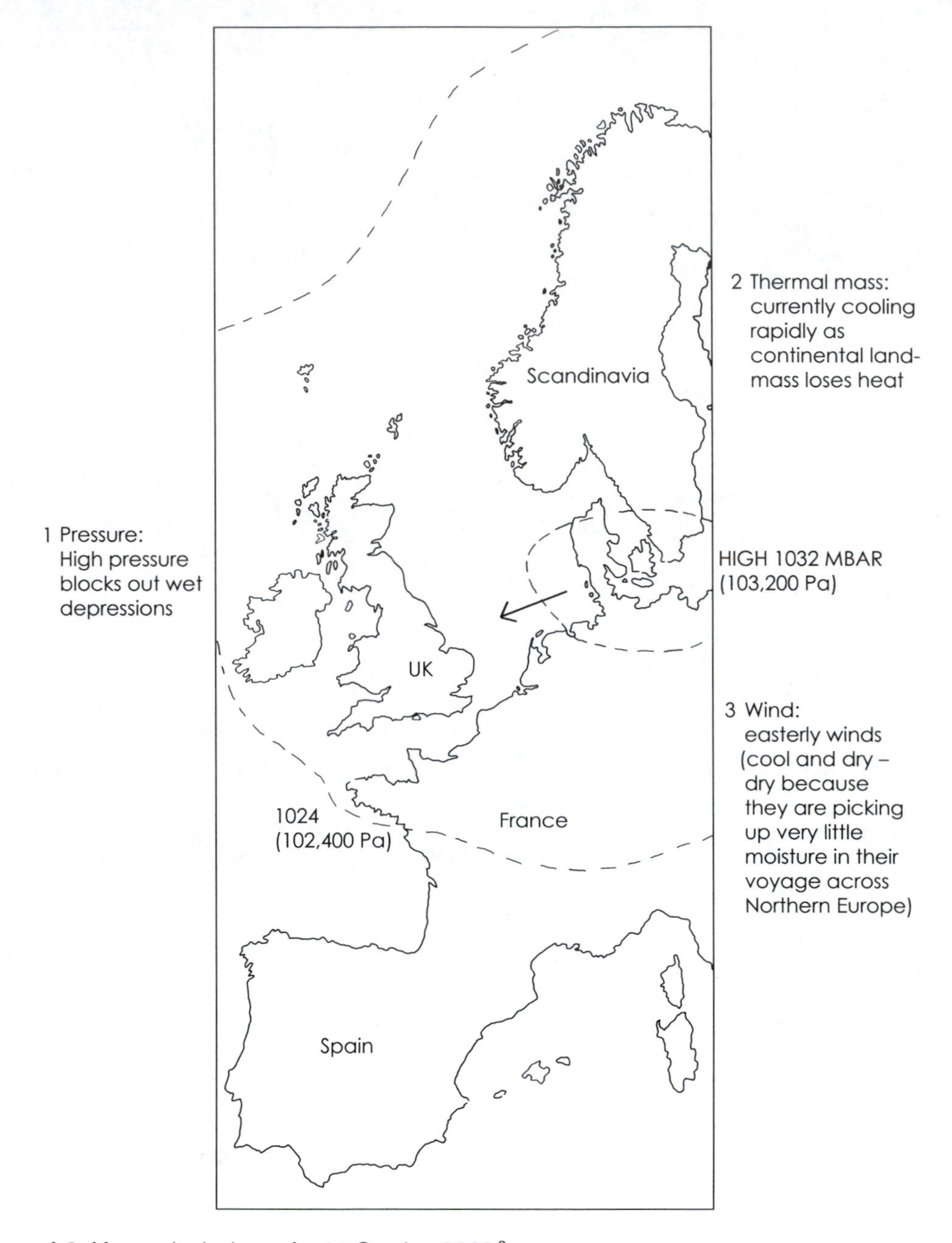

A.3 Meteorological map for 17 October 2003.[8]

Air and ground temperatures

Figure A.4 gives air and ground temperature at Falmouth in 1994.

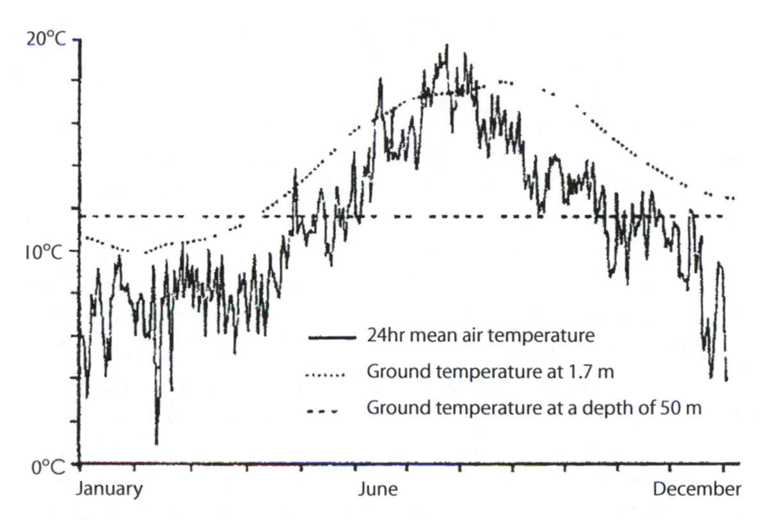

A.4 Temperature data at Falmouth.[9]

Energy and power

Energy is the ability to do work and comes in a variety of forms such as heat and electrical energy – a common unit is the kWh.

Power is the rate of doing work. A tennis player might produce 370W on average during a strenuous match and an electric light bulb might have a power rating of 100 W. If the bulb is left on for ten hours the energy consumed is 1,000Wh or 1kWh.

Form and energy

In the southeast of England on a sunny day in June at noon the incident solar radiation is very roughly 900W per m^2 (the average figure for all hours of daylight is, of course, much lower at about 200W per m^2). This is known technically as the irradiation and is simply a measure of energy over area. During the course of an hour using 900W and a 10m^2 area, about 9kWh will be collected. This is small compared to the situation of a typical motor in a car which might take up very roughly 10m^2 of road but when driven for an hour might consume about 80kWh of energy. What we have done in the past 150 years has been to develop and exploit concentrated sources of energy such as oil and use them in compact pieces of machinery.

As we move towards solar and wind energy we will need to deal with less concentrated (more diffuse) sources and this will inevitably have an effect on our buildings and cities. As suggested in Chapter 4, large areas will be needed as collection surfaces and, to a certain extent (see below), buildings have such surfaces.

Wind energy is similar. The kinetic energy in the wind (which is proportional to the cube of the wind speed hence the great advantage of geographical areas with higher velocities) divided by the area the wind is passing through is the equivalent of the irradiation. In the London area with an average wind speed of about 4m/s, the figure is about 80 W for every square metre the wind passes through. Again, this relatively low figure means that large areas are required to capture this sustainable source of energy.

Solar and wind energy are, of course, variable and that makes it more difficult to use them; however, one of their advantages is that they will be around for long after people have forgotten how to spell oil.

In Chapter 3 it was briefly mentioned that as buildings grow larger the volume increases more than the surface area. The implications for this are varied and require careful examination. On the one hand, for example, this means that there is relatively less heat loss (which is largely a surface phenomenon). A biological example of the implications of this is that, other things being equal, a large bird is less vulnerable in a cold snap than a small one.[10] On the other hand, in buildings and in compact cities the effects of this tend to be that there is proportionately less surface area available for solar energy collection (light and heat) but a proportionately greater volume. This volume is likely to mean more people and, say, a greater need for domestic hot water, more electrical energy for computers, and so forth. The greater volume will also mean a greater heat production (i.e. the 'metabolism' of the organism) and so a requirement for heat dissipation (with this dependent on surface area). The overall result is a need for more sophisticated design for both the surfaces and the servicing systems for buildings and cities.

Another effect on form is that in dealing with the movement of air, in order to reduce the resistance to flow and thus operate within a lower, more natural range of driving forces it is necessary to have larger duct sizes and openings. This can be seen in numerous examples in the book, including the BRE Environmental Building (Chapter 4) and The Contact Theatre (Chapter 6).

Thermal conductivity, thermal resistivity and U-values

The thermal conductivity (W/mK) is a measure of how much heat a material will transmit per unit thickness and unit degree of temperature difference. The lower the value, the better it acts as an insulant. The thermal resistivity is its inverse (mK/W).

Table A.2 gives some typical values.

As can be seen from Table A.2 those materials with very low values of thermal conductivity, i.e. the good insulators, benefit from (stationary) air being a poor conductor of heat.

U-values

The U-value is the rate of heat flow per unit area through a construction element per degree of temperature difference. Very simply, it provides a good idea of the degree of insulation of a real construction. Lower U-values mean better insulation (i.e. lower heat loss). Table A.3 gives some comparative examples.

Thermal mass

Thermal mass and thermal capacity are used somewhat loosely and often interchangeably. Very simply, and in the context of buildings, thermal mass refers to the amount of heat the structure can absorb for a given temperature rise. Thus, spaces

Table A.2 Thermal properties of materials[a]

Material	*Thermal conductivity (W/m K)*	*Thermal resistivity (m K/W)*
Air	0.026	38.5
Mineral wool	0.036	27.8
Hair felt	0.039	25.6
Sheep's wool	0.045	22.22
Compressed strawslab	0.085	11.8
Timber, softwood	0.13	7.7
Compact snow	0.21[c]	4.76
Cardboard	0.22	4.6
Mud (5% dry weight)	0.43	2.3
Water at 20°C	0.6	1.67
Brickwork[b]	0.7–1.4	1.4–0.7
Window glass	1.05	0.95
Concrete[b]	1.4	0.7
Limestone	1.5	0.67
Ice at −20°C	2.45	0.4
Copper	200	0.005

Notes
a All data from note 11 unless otherwise noted.
b Note 12.
c Note 13.

with high thermal mass can absorb more heat before the temperature rises significantly and so are more likely to remain comfortable longer.

The broader scientific concept of thermal mass is based on the specific heat capacity of a material. This is the amount of heat a material can store per unit of mass and per unit of temperature change. Water has a particularly high specific heat capacity of 4,200 J/kg K (which means that it requires 4,200 J of energy to raise 1 l of water by 1°C). This has been particularly advantageous for life (most life forms are mainly water – humans are about 60 per cent water) because it helps guard against the effect of temperature extremes. (Note 15 gives a reference for additional information.)

Force and pressure

Force

The unit of force, the newton (N), is the force to accelerate an object with a mass of 1 kg at 1 m/s^2. The forces in the arms of a Neanderthal hunter are shown in Figure A.5.

Pressure

Pressure is force per unit area. So for example in the illustration above, assuming a target area of about 25 mm × 25 mm in this poor animal (reindeer, elk, horse and bison were the favoured prey),[17] the pressure would be about 1,600,000 Pascals (Pa).

The pressures associated with more natural forces in buildings tend to be much lower as can be seen in Table A.4 which gives a range.

Table A.3 Approximate U-values

U	*W/m² K*
a. Wigglesworth/Till House wall with straw bales	0.16 
b. Roof of 100 mm of sheep's wool in timber frame	0.35 
c. Orkney Croft	1.6 
d. Igloo	1.8 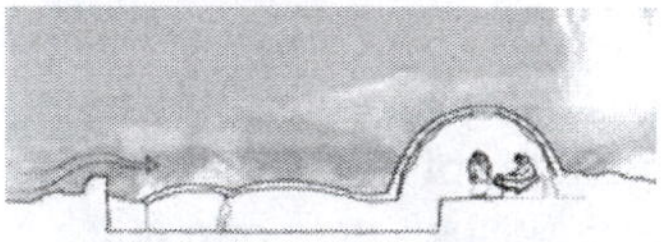
e. Farnsworth House glass wall	5.6 
f. Temporary cardboard shelter mat under bushes	6.7 

Notes

1 As a very approximate comparison, if one could talk about the U-value of a polar bear's fur (assuming a coat of 62 mm) the figure would be about 1.2 W/m^2K.[14]

2 A continental quilt (10.5 togs) has a 'U-value' of about 1 W/m^2K.

3 Enthusiasts of Lewis Carroll will remember that in *Alice's Adventures in Wonderland*, the March Hare's roof was thatched with fur – and the chimneys were shaped like ears.

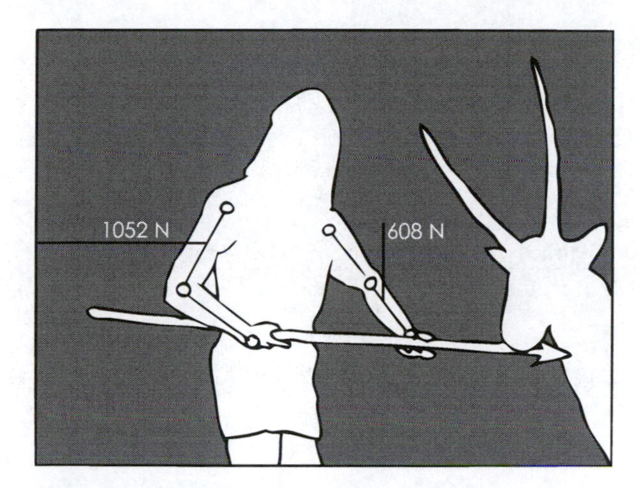

A.5 Forces in a Neanderthal hunter.[16]

Table A.4 Approximate pressures

Item	*Pressure (Pa)*
Blowing on your outstretched hand	0.05
Pressure due to the stack effect for a height of 12 m and a temperature difference of 5 K – see note 1	2.6
Velocity pressure of a 4 m/s wind – see note 2	9.6
1 l of milk on a 1 m^2 table	10
Pressure in the human lung	−150 to +150
Pressure in a mechanical ventilation system	150 to 300
Human blood pressure (systolic)	16,000
Acceptable bearing pressure for foundations on gravel	200,000
Typical compressive strength for concrete	35,000,000

Notes

1 The pressure difference is given approximately by 0.043 × height × temperature difference between inside and outside air.

2 The velocity pressure is given by 0.6 × velocity squared.

Light

The nature of light

Figure A.6 shows the spectrum of electromagnetic radiation with light occupying the range from about 400 to 700 nm. (One nm is 1×10^{-9} of a metre.)

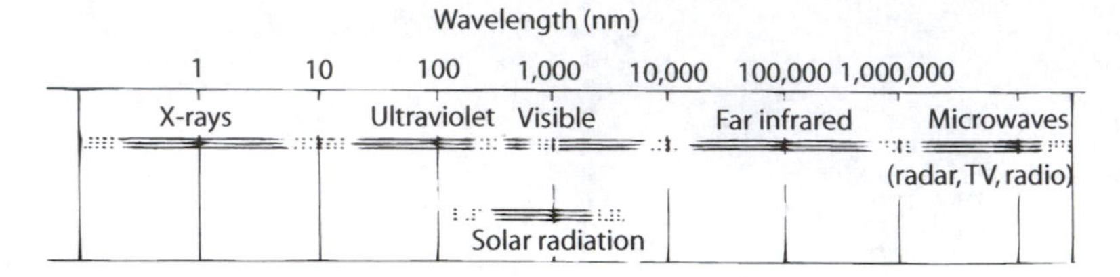

A.6 The electromagnetic spectrum.[18]

The spectral distribution of solar radiation at the Earth's surface is shown in Figure A.7. Note the difference between solar radiation generally and north sky daylight.

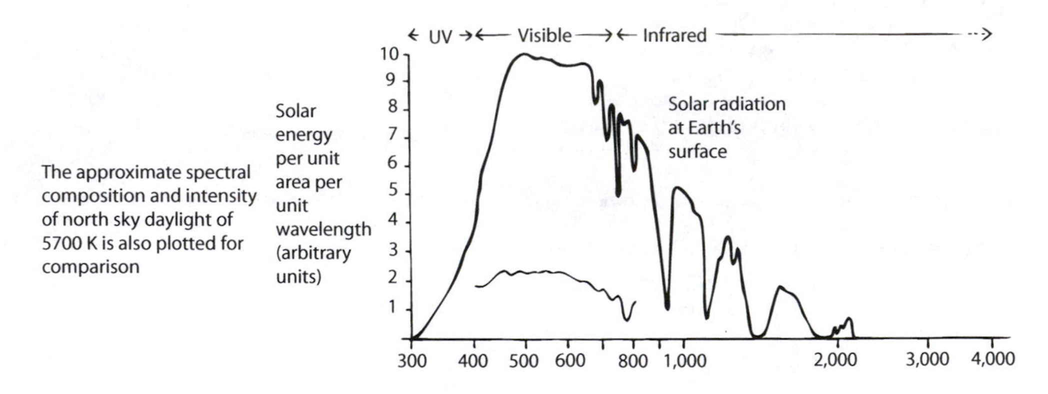

A.7 Spectral distribution of solar radiation and north sky light at the Earth's surface.[19]
Note
Data for north sky daylight is based on a temperature of 5,700 K (infrared not shown).

Recommended light levels[20]

Space	Standard maintained Illuminance (lux)
Homes	
Living rooms	50
General	
Sewing and darning	300
Studies	300
Offices	
General	500
Libraries	
Reading room	300

It should be kept in mind that the indiscriminate use of such numbers can create spaces without feeling. A dimly lit dining room (at 25 lux) can be dismal. Three candles will provide about the same light level at the table with an entirely different effect.

A simple example of daylighting

How does one start to analyse the lighting in Pieter de Hooch's interior (Fig. 7.4)? First, note the position of the windows – they cover the full width of the visible room and run right up to the ceiling, thus allowing light deep into the space. Then, one asks what kind of glazing is it? We can see (easier in the original, of course) a variety of sharp shadows, for example, that of the maid to the right. The woman is holding the glass in the light and one can see its clarity. All of the light clothes of the figures are well illuminated and there is a marked contrast with the darker tones. All of this argues for glass which is transparent. (Such a view is supported by a Vermeer's painting 'A Lady Reading at a Window' which has similar leaded glazing. In that painting the window is open into the room and one can see quite clearly through the glass.) Note, though, that the seated man to the right is looking partly at the woman and partly through the wine and at the window. This suggests that although it is sunny the sun is not directly in his eyes – the angle of the shadows might suggest an elevation of about 50°. (Nothing, of course, requires de Hooch to present the scene completely realistically but verisimilitude is a reasonable starting point.) The walls are painted in a light off-white, the floor appears half white and half grey and the ceiling is light brown timber. Thus, there is reasonable reflectivity and the light is carried deep into the room. So, quantitatively, the ratio of the light on the table to that outside might be about 1:8, giving a daylight factor of about 12.5 per cent. Perhaps it is higher, but a quick mock-up of the setting in a living room and measurements indicate that that is about right. Deeper into the room, say to the right of the maid, it might fall to below 5 per cent. If we guessed that the outside light level was at least 20,000 lux we would have 2,500 lux on the table which would be amply generous and provide real sparkle. Even with an overcast sky one could work comfortably on that table. But as to the poetry of the painting, note that the lower shutters to the rear are closed, thus creating asymmetry. And as to magical realism, the upper shutters are fast against the ceiling with no painted way of maintaining them in that position. (And that is why functionalism alone is an inadequate way of explaining art – or architecture.)

Comfort

Comfort is a subjective matter and will vary with individuals. It involves a large number of variables, some of which are physical with a physiological basis for understanding. Classically, for thermal comfort they include:

- air temperature and temperature gradients
- radiant temperature
- air movement
- ambient water vapour pressure (more commonly and more loosely described as the 'humidity')
- amount of clothing worn by the occupants
- occupants' level of activity.

Other factors influencing general comfort are light levels, the amount of noise and the presence of odours. Individuals are also affected by such psychological factors as having a pleasant view, having some control of their environment and having interesting work. For some variables it is possible to define acceptable ranges, but the optimal value for these will depend on how they interact with each other, e.g. temperature and air speed, and personal preference.

What we know is that humans can become accustomed to a wide range of temperatures and relative humidities. Nonetheless, approximately, in Europe one might say that keeping temperatures in buildings in the range of 18 or 19°C in winter to 27°C to 29°C in summer is broadly comfortable; relative humidities in the approximate range of 40 per cent to 70 per cent should satisfy many people (and certainly more than Stendhal's 'happy few').

Further reading

Thomas, R. (2006) *Environmental Design*, London: Spon Press.

Appendix B
A time line

The time-line shown in Figure B.1 gives a few 'dates' of relevance to the themes of this book.

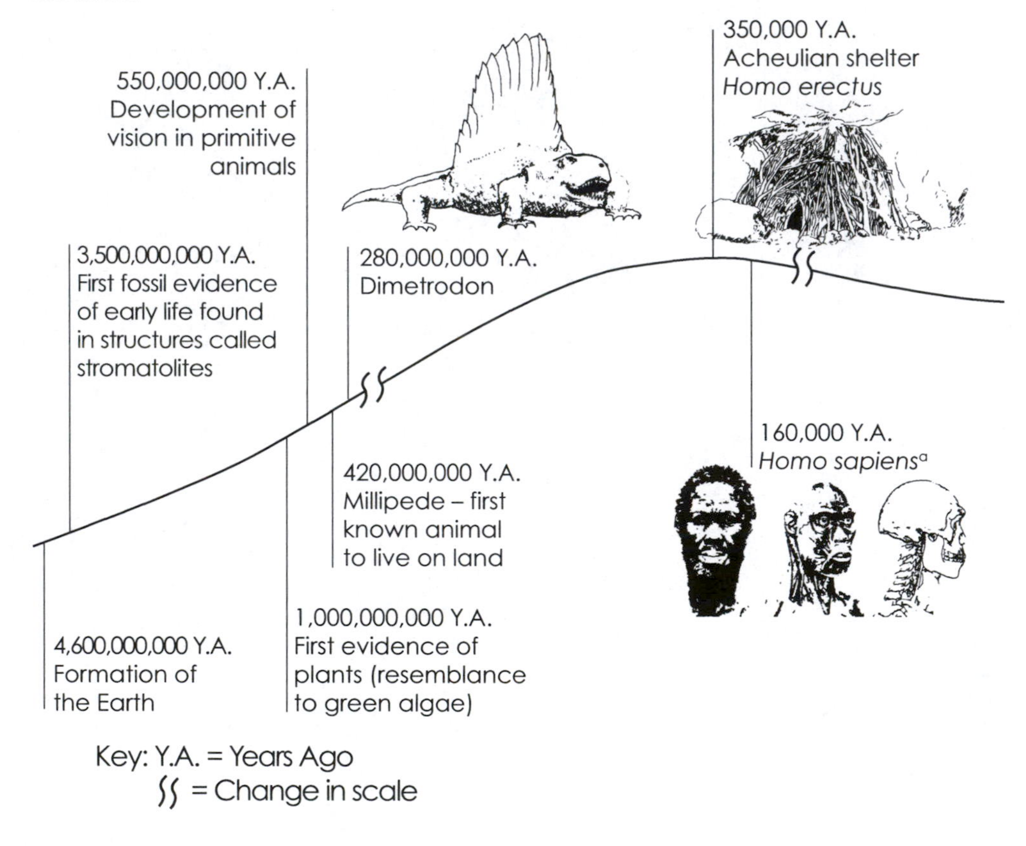

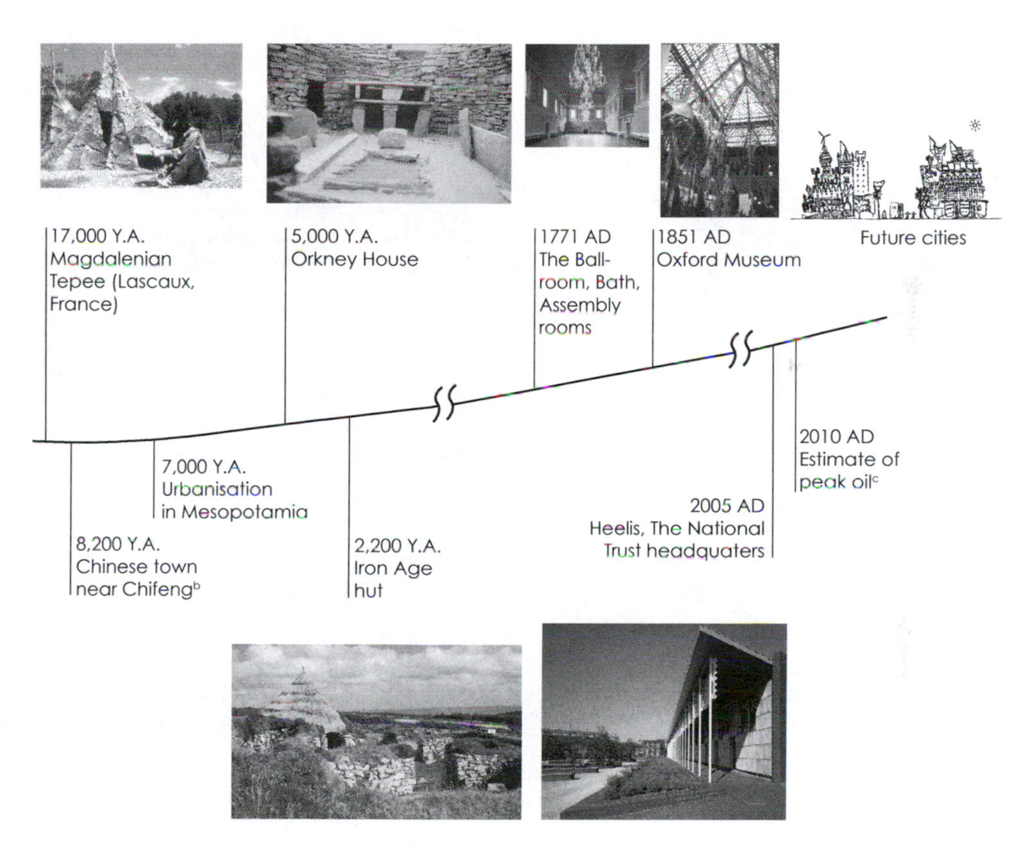

a The illustration shows a reconstruction of Ethiopian fossils of early *Homo sapiens*.[1]

b Near Chifeng (about 500 km north-east of Beijing) the remains of substantial urban dwellings from a 'very majestic' town estimated to be about 8,200 years old have been found.[2]

c 'Peak oil' refers to the point when oil production reaches its peak – note that this is very different from when oil runs out given that this is likely to be a protracted (and painful) process. One specialist, Colin Campbell, estimates that 'the peak of all kinds of oil together ... is 2010'.[3]

B.1 A time line.

Notes

Preface

1 Garnham, T. (1999) 'Building – Alan Short in Manchester', *Architecture Today*, June, pp. 24–32.
2 Although written more than 40 years ago, C.P. Snow's distinction between what he called 'The Two Cultures' – science and the humanities – continues to resonate for some (and reared its head in the writing of this book). Snow felt that he was the only person amongst his fellow academics at Cambridge who could talk freely with colleagues from both sides.
3 William of Occam (1300?–1350?) was an English philosopher. The maxim 'Entities are not to be multiplied without necessity', although not stated explicitly as such in any of his works, has become known as 'Occam's razor'.
4 The retreating Arctic polar ice is affecting the bears' ability to hunt for seals. This in turn affects body weight and fertility. For these and a number of other reasons it is feared that polar bears could be only decades away from extinction. Connor, S. (2004) 'Meltdown: Arctic wildlife is on the brink of catastrophe', *Independent*, 11 November, p. 34.

1 Introduction

1 Wolpert, L. (1992) *The Unnatural Nature of Science*, London: Faber & Faber, pp. 85–87.
2 *Oxford English Dictionary* (1978), Oxford: Oxford University Press.
3 Frampton, K. (2002) 'Towards a Critical Regionalism', in *Labour, Work and Architecture*, London: Phaidon, p. 86.
4 In TG's view Buckminster Fuller's 'Spaceship Earth' and James Lovelock's 'Gaia' became two of the most striking metaphors coined in the twentieth century for describing mankind's relationship to the planet. Lovelock's science is considered suspect by some for its overarching conclusions. Nevertheless, his conception of Gaia as the seemingly balanced, complex inter-relationship of all biological life on Earth and its importance for maintaining a relatively stable composition of the atmosphere strikes a chord with many concerned at environmental degradation caused by mankind's ever-increasing technological capacity.

2 Site and setting

1 'Most living things are plants – roughly 95 per cent of the world's living matter in fact. Plant biodiversity underpins the entire world and we neglect it at our peril.' Professor John Parker, 'Gatsby Foundation to Fund Institute of Global Evolution', *Cam*, no. 44, p. 205.
2 The Garnham family tree, for example, shows the author's particular branch moving no more than four or five villages, about 12 miles, between 1591 and 1952.
3 Tudge, C. (1991) *Global Ecology*, London: Natural History Museum Publications, pp. 61–62. Much of the environmental material in this chapter is drawn from this book.
4 Although most structures would have been made of branches, archaeologists have found shelters made from mammoth bones and tusks in Siberia. Thomas, H. (1995) *The First Humans: The Search for our Origins*, London: Thames and Hudson, p. 118.
5 Frankl, G. (1992) *Archaeology of Mind*, London: Open Gate Press, pp. 60–61.
6 Muir, E. (1993) *An Autobiography*, Edinburgh: Canongate Press, pp. 20–21.
7 Hawkes, J. (1963) *Prehistory*, London: George Allen & Unwin, p. 158.
8 Tudge, op. cit., p. 44.

9 Rapoport, A. (1969) *House Form and Culture*, New Jersey: Prentice Hall, pp. 98–99.
10 Oliver, P. (2003) *Dwellings: The Vernacular House Worldwide*, London: Phaidon, pp. 23–24. Brunelleschi achieved fame as a 'father' of the Renaissance partly because of building the Duomo at Florence, using a similar principle of self-supporting construction in brick and stone.
11 Imbert, B. (1992) *North Pole, South Pole: Journeys to the Ends of the Earth*, London: Thames and Hudson, pp. 57, 79–100.
12 Ibid., pp. 90–91.
13 Hill, J. and Woodland, W. (2003) 'Subterranean settlements in southern Tunisia; environmental and cultural controls on morphology, community dynamics and sustainability', *Geography*, vol. 88, part 1, pp. 23–39.
14 Oliver, P. (ed.) (1977) *The Encyclopaedia of Vernacular Architecture*, vol. 2, Cambridge: Cambridge University Press, p. 875.
15 Mithen, S. (1998) *The Prehistory of Mind*, London: Phoenix, pp. 222 and 256.
16 Darvill, T. and Thomas, J. (eds) (1996) *Neolithic Houses in North-west Europe and Beyond*, Oxford: Oxbow Books. Hodder, I. (1990) *The Domestication of Europe*, Oxford: Blackwell.
17 Hoskins, W.G. (1971) *The Making of the English Landscape*, Harmondsworth: Penguin Books, pp. 17–32.
18 Renfrew, C. (ed.) (1990) *The Prehistory of Orkney*, Edinburgh: Edinburgh University Press, pp. 23–27. The larger deciduous trees of mainland Britain did not establish themselves this far north (Orkney is on latitude 59°, the same as Hudson Bay and further north than Moscow).
19 Hawkes, op. cit., pp. 137–37.
20 Some of the houses at Skara Brae have been washed away by the encroaching Atlantic.
21 Oliver, op. cit., pp. 842–848 and p. 993.
22 See Garnham, T. (2004) *Lines on the Landscape, Circles from the Sky*, Stroud: Tempus, for more details of Neolithic Orkney.
23 Rapoport, op. cit., p. 42.
24 Ibid., pp. 42–43. Language, song, dance and ritual may have been more advanced amongst early humans.
25 Ibid., p. 84.
26 See, for example, her poem 'The Monument', which includes the lines:

> It is an artifact
> of wood. Wood holds together better
> than sea or cloud or sand could by itself,
> much better than real sea or sand or cloud.
> It chose that way to grow and not to move.

Seamus Heaney dicusses this and other poems of Elizabeth Bishop in 'Counting to a Hundred' (2002) *Finders Keepers: Selected Prose 1971–2001*, London: Faber & Faber, pp. 332–346.
27 Bachelard, G. (1969) *The Poetics of Space*, Boston: Beacon Press, p. 7.
28 Heidegger, M. (1977) *The Question Concerning Technology*, New York: Harper and Row, pp. 10–14.
29 Fathy, H. (1986) *Natural Energy and Vernacular Architecture: Principles and Examples with Reference to Hot Arid Climates*, Chicago: University of Chicago Press, p. 3. R.T. does not agree with Fathy's view expressed here.
30 Torture was, of course, rife at this time as a method to secure the truth. Bacon advocated, amongst other things, the experimental method to find out the workings of nature that would be a good thing in itself and also of benefit to humankind. Bacon may be considered to be the founding father of the science of refrigeration and paid a tragic price for his experimental curiosity when he died of a chill caught when stuffing a chicken with snow.
31 *BRE Digest* 350: Climate & Site Development.
32 *Guardian*, 'Environment', 13 July 2005, p. 12.
33 *New Scientist*, 21 January 2006, p. 36.

3 Building design 1: smaller buildings and the creation of environments

1 Wigglesworth, V.B. (1966) *The Life of Insects*, New York: The New American Library, p. 18.
2 Ibid., p. 27.
3 Thomas, R. (2000) *Environmental Design*, London: E. & F.N. Spon, p. 88; Mithen, op. cit., pp. 18–23.
4 Mellart, J. (1967) *Çatal Hüyük*, London: Thames and Hudson.
5 The earliest houses in Orkney were the same size and shape as prehistoric boats and their precursors in the Western Isles and Ireland were made of a woven, basket-like construction. For more on this, see Garnham, op. cit., pp. 32–53.
6 Bachelard, op. cit., p. 17.
7 Ibid., p. 33.
8 Eliade, M. (1971) *The Myth of the Eternal Return*, Princeton: Princeton University Press.
9 Eyck, Aldo van (1960) 'The Interior of Time', in G. Baird and C. Jencks (eds) *Meaning in Architecture*, London: The Cresset Press, pp. 170–213.
10 Campbell, J. (1990) 'Bios and Mythos', in *The Flight of the Wild Gander*, New York: Harper Perennial, pp. 52–55.
11 Ibid.
12 Redfield, R. (1968) *The Primitive World and its Transformation*, London: Penguin, p. 111.
13 Marzluff, J.M. and Angell, T. (2005) *In the Company of Crows and Ravens*, New Haven: Yale University Press, pp. 12–13.
14 Rapoport, op. cit., pp. 18–45.
15 Elvin, M. (2004) *The Retreat of the Elephants: An Environmental History of China*, New Haven: Yale University Press.
16 Fathy, op. cit., p. xxi.
17 Hamblyn, R. (2002) *The Invention of Clouds*, London: Picador, p. 16.
18 Eliade, M. (1959) *The Sacred & the Profane*, New York: Harcourt, Brace, Jovanovich, pp. 163–164; Ewart Evans, G. *Ask the Fellows who cut the Hay*, London: Faber & Faber.
19 Vitruvius (1960) *The Ten Books on Architecture*, New York: Dover, pp. 38–41.
20 Rykwert, J. (1972) *On Adam's House in Paradise: The Idea of the Primitive Hut in Architectural History*, New York: The Museum of Modern Art.
21 Schildt, G. (ed.) (1998) *Alvar Aalto in His Own Words*, New York: Rizzoli, p. 117.
22 Ibid., pp. 118–119.
23 Ibid., p. 119.
24 Blackburn, S. (1994) *The Oxford Dictionary of Philosophy*, Oxford: Oxford University Press, p. 101.
25 Heidegger, op. cit., pp. xxv–xxxii. This interpretation developed by Heidegger is not accepted by all, including R.T.
26 Lloyd, N. (1976) *The History of the English House*, London: Architectural Press, p. 7.
27 Snow, C.P. (1951) *The Masters*, London: Penguin, p.11.
28 *Eco Tech. Sustainable Architecture Today*, July 2004, pp. 12–13.
29 Smith, P. (2001) *Architecture in a Climate of Change*, Oxford: Butterworth-Heineman, pp. 60–64.
30 Lloyd, op. cit., p. 30.
31 Batty Langley, *The Builder's Jewel*, *The Builder's Chest Book* and *The Builder's and Workman's Treasury of Designs.*
32 Lloyd, op. cit., p. 164.
33 Ibid., pp. 2–3. The similarity between the Medieval and Saxon houses is also seen in the derivation of the word hall from the Old Saxon *halla*.
34 Dunn, M. (1999) 'Japan', in G. Fahr-Becker, *The Art of East Asia*, Cologne: Köneman.
35 Lethaby, W. (1979) *Philip Webb and His Work*, London: Raven Oak Press, p. 250. (These phrases are in an appendix of the second edition. Originally published by Oxford University Press, 1935.)
36 For more on this, see Garnham, T. (1993) *Melsetter House*, London: Phaidon.
37 Muthesius was appointed to the German Embassy in London from 1896 to 1903, at first as a 'Technical Attaché' and later as 'Cultural Attaché'. Lethaby was almost certainly his guide to the latest developments in English architecture.

38 Scully, V. (1971) *The Shingle and the Stick Style*, New Haven: Yale University Press.
39 Eliade, M. *The Sacred & the Profane*, op. cit., pp. 56–57.
40 Drew, P. (1985) *Leaves of Iron*, Sydney: The Law Book Co., pp. 49–72.

4 Building design 2: the environments of larger buildings

1 Garnham, T. (1992) *The Oxford Museum*, London: Phaidon.
2 Gould, Stephen Jay (1978) 'Size and shape', in *Ever since Darwin: Reflections in Natural History*, London: Penguin, p. 177.
3 Ibid., p. 175.
4 Giedion, S. (1948) *Mechanization takes Command*, Oxford: Oxford University Press, p. 556.
5 Banham, R. (1969) *The Architecture of the Well-Tempered Environment*, London: Architectural Press, p. 181.
6 Ibid., p. 224.
7 Liebel, B. and Brodwick, J. 'Choosing the right light', in *ASHRAE Journal*, December 2005, p. 122.
8 Le Corbusier (1946) *Towards a New Architecture*, London: Architectural Press.
9 Ibid., pp. 171–172.
10 Brace Taylor, B. (1987) *Le Corbusier: The City of Refuge Paris*, Chicago: Chicago University Press, p. 80.
11 Ibid., pp. 112–117.
12 R.T. has a different view, for the concept of alien is quite complex and could be applied to numerous examples of architecture in the past where new ideas, concepts and forms were explored.
13 R.T.'s opinion is that the lesson from biology is not that things have to be 'open' or that they can not be 'sealed'. It is that they must be adapted to their surroundings. As it turns out, most life forms are 'open' (as indeed are most buildings) to the extent that they need to obtain food and dispel wastes. What we need to do is shift our buildings from a reliance on fossil fuels.
14 Sennett, R. (1990) *The Conscience of the Eye: the Design and Social Life of Cities*, London and Boston: Faber & Faber, p. 52.
15 Heidegger, M. op. cit., pp. xxx–xxxix and 19–35.
16 Banham, R. (1960) *Theory and Design in the First Machine Age*, London: Architectural Press, pp. 19–35.
17 Banham, op. cit., p. 265.
18 Ibid., pp. 86–90 and pp. 246–252.
19 Ibid., p. 256.
20 R.T. doesn't share this view. A building's multiple roles of, for example, providing enjoyable spaces for people – and in the case of the Pompidou Centre, showcasing art – continue. In the case of the Pompidou Centre – a building for modern art – the symbolic role of architecture as a demonstration of modernity for Paris and France was also important.
21 Le Corbusier (1989) *Le Poème de l'Angle Droit*, Paris: Le Corbusier Foundation, p. 69.
22 Banham, op. cit., p. 158.
23 Farmer, J. 'Battered bunkers', *Architectural Review*, January 1987, pp. 60–65.
24 See, for example, Battle, G. and McCarthy, C. (2001) *Sustainable Ecosystems and the Built Environment*, Chichester: Wiley-Academy.
25 Research by Watkins *et al.*, 2002, referred to by Alan Short, *Architecture Today 167*, April 2006, p. 54. The account given here of the School of Slavonic and East European Studies is based on the articles in this issue of *Architecture Today*, pp. 48–56.
26 Banham, op. cit., p. 21.
27 Ibid., pp. 20–21.
28 Ibid., pp. 280–286.
29 Dennis, M. (1988) *Court and Garden: From the French Hôtel to the City of Modern Architecture*, Boston: MIT Press, pp. 215–219.
30 It has been said that the Paris Hilton proves Darwin was wrong. 'How could a process of "survival of the fittest" ever produce something that was so completely and utterly useless – only a designer could do that.' Bill Mitchell, *RIBA Journal*, vol. 112, no. 12, December

2005. This comment reiterates the difference between nature and culture discussed earlier here.

31 Theodoulou, M. 'The first beach BBQ – 12,000 years ago', *The Times*, 23 November 2005, p. 37.

32 Emma Graham-Harrison, *Guardian*, 30 September 2005, 'China ponders cost of energy-guzzling industries', *Daily Yomiuri*, p. 20.

33 Paulina Wojchiechowska (2001) *Building with Earth*, White River Junction, VT: Chelsea Green Publishing.

34 *Guardian*, 26 November 2005, p. 8. Perhaps a more detailed examination of the data would indicate that when the viewers were architects the response was reversed.

35 Morse, E. (1961) *Japanese Houses and their Surroundings*, New York: Dover, p. 127.

36 Frampton, op. cit.

37 Heaney, Seamus (2002) *Finders Keepers*, London: Faber & Faber, pp. 77–95.

5 Heating, cooling and power

1 Chapman, J. (2003) 'Apes that blow their families a bedtime kiss', *Daily Mail*, 3 January, p. 19.

2 In the mantle of the Earth rocks in a fluid state due to high temperature and pressure move slowly in convection currents which drive plate tectonics. As the movement of plates is largely responsible for the geology of the Earth's surface (in conjunction with phenomena in the lithosphere and atmosphere) and as our lives and livelihoods are so dependent on geology, R. Fortey has used the happy phrase that we 'may all ultimately be the children of convection' (Fortey, R. (2004) *The Earth – An Intimate History*, London: Harper Perennial, p. 429).

3 The history of architecture has many examples of architects who were masters of the arts and sciences but one which is particularly appropriate here is that of Claude Perrault. The east façade of the Louvre is generally attributed to him but it is less well known that he was a distinguished physician who died from a 'cut incautiously made while dissecting the cadaver of a camel which had died in the Jardin du Roi' in Paris (see p. 31 of J. Rykwert's (1983) *The First Moderns*, Boston, MA: MIT Press).

4 Schmidt-Nielsen, K. (1988) *How Animals Work*, Cambridge: Cambridge University Press.

5 We are indebted to R. Yagilis (1985) *The Desert Camel*, Basel: Karger, for the principal facts in this discussion.

6 The Museum of Manchester, University of Manchester, geology display, 2003.

7 We would like to thank Arups for some of the information on the thermal characteristics of igloos.

8 Fitch, J.M. (1976) *American Building 2. The Environmental Forces That Shape It*, New York: Schocken.

9 Brundrett, G.W. (1973) Thermal Comfort in Buildings, Building Society for Engineering Services, February meeting.

10 Anon. (2001) *ASHRAE Handbook – Fundamentals*, Atlanta: ASHRAE, p. 8.1.

11 Anon. (2002) *ASHRAE Handbook – Fundamentals*, Atlanta: ASHRAE, p. 9.11.

12 O'Callaghan, P.W. (1978) *Building for Energy Conservation*, Oxford: Pergamon.

13 Anon. (2001) *ASHRAE Handbook – Fundamentals*, Atlanta: ASHRAE, p. 8.1.

14 Benton, M. (1995) *Vertebrate Palaeontology*, London: Chapman and Hall.

15 Settles, G. (1997) 'Visualizing full-scale ventilation airflows', *ASHRAE Journal*, July, pp. 19–22.

16 Assumed U-values in W/m^2K: 0.15, roof; 0.15, walls; 0.3, floor; 1.5, glazing.

17 Clifton-Taylor, A. (1972) *The Pattern of English Building*, London: Faber & Faber, pp. 259–260.

18 A note on time and energy: before the shift to fossil fuels human societies relied on solar energy that had only recently been radiated to the Earth. Energy-harvesting plants fed animals and both fed humans. Combustion of wood and oils of vegetable or animal origin and moving air and water to drive mills and pumps were all based on what we might call 'new' solar energy. The shift to fossil fuels was a shift to 'old' solar energy. For example, coal which until the early nineteenth century was consumed in negligible amounts, and a great deal of coal is found in sedimentary rocks known as Coal Measures which are

300–315 million years old. (For a further discussion see Singer, S. (1970) 'Human energy production as a process in the biosphere', in *The Biosphere*, San Francisco: Scientific American.)
19 Billington, N. and Roberts, B. (1982) *Building Services Engineering*, Oxford: Pergamon, p. 144.
20 Cited in Banham, R. (1969) *The Architecture of the Well-tempered Environment*, London: The Architectural Press, p. 109.
21 Radiators developed in the first half of the nineteenth century as steam and hot water heating were introduced. A very early example was a crude box of tin plate used as a radiator by James Watt to heat his writing room in 1784 (cited in Billington and Robert's *Building Services Engineering*, p. 101).
22 Butti, K. and Perlin, J. (1981) *A Golden Thread*, London: Marion Boyars.
23 The estimated energy consumption for the building was 38 kWh/m^2 – y for heating and hot water and 45 kWh/m^2 – y for all electricity. This was based on a 30 per cent improvement on best practice in the early 1990s.
24 Behling, S. and Behling, S. (2000) *Solar Power*, London: Prestel.
25 R. Mulvey (2004) Private communication.
26 See note 20, p. 104.
27 We are indebted to W.J Broad's article 'How dolphins keep their cool in the zone that counts' (*New York Times*, 19 January, 2002, p. 25) for the basis of our discussion.
28 Pierce, M.A. (1995) 'A history of cogeneration', *ASHRAE Journal*, May, pp. 53–59.
29 Historically we have been moving towards fuel sources which have lower carbon contents and so the logical environmental move is towards the no-carbon, non-polluting fuel, hydrogen. For example, for a given quantity of energy provided, wood has about 3 units of carbon, oil has 2 and gas has 1.5. Hydrogen is likely to be produced using renewable energy sources, distributed via pipelines and used in fuel cells.

6 Ventilation

1 Plester, J. (2003) 'Weather eye', *The Times*, 21 October.
2 L'Homme, J.C. (2004) *Les Energies Renouvelables*, Paris: Delachaux et Niestle.
3 Anon. (2004) 'How the honeybees do it', *ASHRAE Journal*, August, p. 11.
4 Slezec, A.-M., Janvier, P. and Van Praet, M. (eds) (1991) *On a Marché Sur la Terre*, Paris: Editions ICS/Museum Nationale d'Histoire Naturelle.
5 Bahadori, M. (1978) 'Passive cooling systems in Iranian architecture', *Scientific American*, 238: 2, 144–154.
6 Conventionally, the Iron Age in Britain is taken to end with the Roman conquest.
7 Olly, J. (1985) 'The Reform Club', *The Architect's Journal*, 27 February, 34–60.
8 Swenarton, M. (1993) 'Building – Low energy Gothic: Alan Short and Brian Ford at Leicester', *Architecture Today*, 41: 20–30.
9 Thomas, R. (ed.) (2005) *Environmental Design*, London: E. & F.N. Spon.
10 Rickaby, P. (1991) 'The art of energy: Peake Short & Partners in Malta', *Architecture Today*, 14: 34–40.
11 Anon. (1986) – *Encyclopedia Britannica*, vol. 26, p. 787.
12 Garnham, T. (1999) 'Building – Alan Short in Manchester', *Architecture Today*, vol. 99: 24–39.
13 Quincey, R., Knowles, N. and Thomas, R. (1997) The Design of Assisted Naturally Ventilated Theatres. CIBSE National Conference Proceedings.
14 Thomas, R. (ed.) (2005) op. cit., pp. 222–230.
15 Koralek, P. (2005) 'Building – On Track: FCB in Swindon', *Architecture Today*, vol. 161: 66–77.

7 Light and shade

1 Genesis 1–2.
2 Hawkes, J. (1962) *Man and the Sun*, London: The Cresset Press, p. 47.
3 Tudge, op. cit., pp. 12–15. A related theory is that relatively simple compounds formed in this way were absorbed in clay, which acted as a catalyst causing them to react together.

Other theories are that carbon-based life was preceded by 'living' clay, based on silicone, or that organic molecules arose elsewhere in the universe and arrived on Earth as stardust.

4 *European Directory of Sustainable and Energy Efficient Buildings*, 99, p. 112.
5 Hopkinson, R.G., Etheridge, P. and Longmore, J. (1966) *Daylighting*, London: William Heineman, p. 5.
6 Ruspoli, M. (1987) *The Caves of Lascaux*, London: Thames and Hudson.
7 Messham, S. (1978) *Gas An Energy Industry*, London: HMSO.
8 Ibid.
9 Harkness, E.L. and Mehts, M.L. (1978) *Solar Radiation Control in Buildings*, London: Applied Science, p. 118.
10 Soane bought No. 12 Lincoln's Inn Fields in 1792 and acquired No. 13 in 1810 and No. 14 in 1823. He continued to develop the house and turned it into a museum throughout the rest of his life.
11 Craddock, Nigel (1995) Light in the Work of Sir John Soane, unpublished M.Phil., Cambridge University, pp. 51–52.
12 Ibid., pp. 56–57.
13 Bolton, A.T. (ed.) Lectures on Architecture, 1809–1836, Lecture No. VII, p. 118.
14 Craddock, op. cit., pp. 25–30.
15 Ibid., pp. 9–15.
16 Hix, J. (1996) *The Glasshouse*, London: Phaidon, pp. 9–10.
17 Ibid., pp. 23–32.
18 Le Corbusier (1991) *Precisions*, Cambridge, MA: MIT Press, p. 38.
19 Ibid., pp. 53–56.
20 Eliade, op. cit.
21 Simson, O. von (1962) *The Gothic Cathedral*, New York: The Bollingen Foundation, Pantheon Books, pp. 21–58. The Abbot Suger had set his monks at St Denis on the outskirts of Paris to reconcile the newly discovered writings of Plato with the Bible, and 'measure and light' were identified as principal common themes.
22 Brownlee, D.B. and De Long, D.G. (1993) *Louis I. Kahn: In the Realm of Architecture*, New York: Rizzoli, p. 128.
23 Lobell, J. (1979) *Between Silence and Light*, Boulder: Shambala, p. 20.
24 *Architectural Review*, June 1974, p. 333.
25 Frampton, op. cit., p. 82.
26 Ibid., p. 87.
27 Norberg-Shultz, C. (1980) *Genius Loci: Towards a Phenomenology of Architecture*, London: Academy Editions.
28 Pascal, A. 'Molecular basis of photoprotection and control of photosynthetic light-harvesting', *Nature*, vol. 436: 134–137.

8 Cities

1 Rykwert, J. (1976) *The Idea of a Town*, London: Faber & Faber, pp. 27–30. Roman cities were created by a process of biological symbolism. The City of Lougdunum, now Lyon, for example, was founded in 43 BC by the Governor of Gaul who on 9 October, facing the rising sun, traced with a plough drawn by a bull and heifer, the two main roads and then the city's perimeter.
2 Girardet, H. (1992) *The Gaia Atlas of Cities*, London: Gaia Books Ltd.
3 Dennis, M. (1998) *Court and Garden*, pp. 215–216.
4 Girardet, H. (1995) 'Greening society', in Papanek, V. (ed.) *The Green Imperative*, London: Thames and Hudson, p. 18.
5 Girardet, op. cit., p. 20.
6 Ibid., pp. 86–87.
7 Rogers, R. (1997) *Cities for a Small Planet*, London: Faber & Faber.
8 Reader, J. (2004) *Cities*, London: William Heineman, p. 11.
9 Ibid., p. 29. The Sumerian term for garden – *ki-sum-ma* – literally means the place of onions.
10 Morris, A.E. (1994) *A History of Urban Form*, Harlow: Longman Scientific and Technical, pp. 8–10.

11 Ibid., p. 11.
12 Ibid., pp. 41–42.
13 Ibid., p. 43. Often credited with being the earliest planned cities, these are, most probably, those of the Harappan civilisation that arose beside the Ganges in India dating from about 1750 BC.
14 Ibid.
15 *Encyclopaedia Britannica* (2003), Chicago/London: Encyclopaedia Britannica Inc., vol. 8, p. 612.
16 Giradet, *The Gaia Atlas of Cities*, op. cit., p. 117.
17 Rykwert, op. cit., p. 134.
18

> Feng Shui is a system of geomancy which aims to 'harness the natural elements in harmonious ways to create an auspicious environment' (very similar to the approach of the current authors). 'Each element corresponds to a cardinal direction in order of auspiciousness – they are south (fire), east (wood), west (gold) and north (water) while the centre is the earth'. The Forbidden City of Beijing was designed following such principles and the 'city laid out according to the human body: its outer gates were meant to correspond to the head, feet and hands of the body while the maze-like series of internal courtyards housing the emperors represented the intestines'.
>
> (Greenberg, S. (2005) *Enter the Dragon*, RA, Winter, p. 56).

19 Redrawn from Spiller, J. (1973) *Paul Klee Notebooks*, Vol. 1, *The Thinking Eye*, London: Lund Humphries, p. 12; Excerpt from pen-and-ink drawing: City of Cathedrals, 1927/08.
20 Morris, op. cit., p. 99.
21 Ibid.
22 Carl, P. (2000) 'Urban density and block metabolism', in *Architecture, City, Environment*, Proceedings of PLEA, Cambridge, p. 343.
23 Le Corbusier (1987) *City of Tomorrow*, London: Architectural Press, pp. 5–12.
24 Swenarton, M. *Site Planning and Rationality in the Modern Movement*, AA Files 4 (July 1983), pp. 49–59.
25 Short, A. 'Sustainable design in an urban context; three case studies', in Thomas, R. (2003) *Sustainable Urban Design*, op. cit., p. 157.
26 The Modern Movement's interest in sun, light and air arose in part from an interest in health. In Aalto's Paimo Sanatorium the treatment for tuberculosis was isolation from urban pollution in natural surroundings with optimum exposure to sun and fresh air. Le Corbusier's Radiant City extolled the virtues of sun, space and 'greenery' as a remedy to such problems.
27 Redrawn from Le Corbusier's *City of Tomorrow*, op. cit.
28 Steemers, K., Ramos, M. and Sinou, M. (2004) 'Urban diversity', in Steemers, K. and Steane, M.A. (eds) *Environmental Diversity in Architecture*, Spon Press.
29 Collins, G.R. and Collins, C.C. (1986) *Camillo Sitte: The Birth of Modern City Planning*, New York: Rizzoli.
30 Aurigemma, G. (1979) *Giovan Battista Nolli, Architectural Design*, vol. 49, nos 2–4, pp. 27–29.
31 Thomas, R. op. cit., pp. 147–156.
32 Ibid., pp. 167–182.
33 Ibid., pp. 183–188.
34 Rogers, op. cit., pp. 20–22.
35 Carl, op. cit., p. 344.
36 Ibid., p. 345.
37 London energy consumption (all users): $154{,}400 \times 10^6$ kWh; land area: $1.75 \times 10^9\,m^2$. Reference: Anon. (2002) City Limits. A resource flow and ecological footprint of Greater London. Oxford: Best Foot Forward.
38 To cite but one example of the dangers, part of the electricity supply for the city of New York comes from a nuclear power plant some 50 km north on the Hudson River. Residents in an extensive area surrounding the plant have been given masks to wear on days of particular 'terrorist alert' to 'protect' them in the event of an attack on the plant.
39 For a more detailed discussion of these considerations see R. Thomas (ed.) *Sustainable Urban Design*.

40 The reduction of (fossil-filled) energy use is one of the most urgent problems facing cities. One of the more optimistic aspects of this is that it is also often the most economic way of reducing global warming. And in addition there is evidence that it is a more economical proposal than the provision of nuclear power with its attendant dangers. There is a view that it is cheaper, for example, to save energy than to run nuclear power stations (see Anon. 2006) Cheaper to Save than to Run Nuclear Stations, Energy in Buildings and Industry, February, p. 6. For detailed information on the costs of running nuclear power plants see Thomas, S. (2005) The Economics of Nuclear Power: Analysis of Recent Studies, July, PSIRU, University of Greenwich. For data on the cost of energy-saving programmes, see www.nao.org.uk/publications/nao_reports/03–04/0304878.pdf.
41 Anon. (2004) Sustainable Urban Design and Climate, www.bam.gov.au/climate/environ/design/design_a.shtml. 27 October.
42 The first use of any agricultural product should probably be a high-value product (food, paper, other manufactured goods) with the residue being used for energy, where appropriate, and minerals being returned to the land.
43 The pipes shown date from the Soviet era and run from the coal-fired power station to all areas of the city; one is an 'artery' carrying water at 110°C and the other is a 'vein' running at 90°C. The hot water is a by-product of electricity generation at the power station which is thus a major CHP (combined heat and power) plant. The pipes which date from the Soviet era were run above ground to reduce costs and because of difficult ground conditions – the soil is both corrosive and can freeze down to 2.5 m in the coldest winters.
44 'The ecological footprint of a region or community can be viewed as the bioproductive area (land and sea) that would be required to sustainably maintain a region or community's current consumption, using prevailing technology.' See note 37, p. 45. It is a measure of the area required to produce the resources consumed and to absorb the wastes generated. (Since it is based on prevailing technology, it has the limitation of being a constantly varying figure.)
45 Ibid.
46 To give this a human dimension, in the 2004 Olympics in Athens in August the Marathon contestants including the acclaimed athlete Paula Radcliffe were running in 35°C air temperature on tarmac at 49°C (*Guardian*, 23 August, 2004, p. 1). The potential for using these black and thus highly absorbent solar collectors known as roads remains to be realised.
47 Bachelard, op. cit.
48 For but one example among many, in Freiburg, the tramlines have grass in between them. The grass is part of a sustainable urban drainage system, makes the streets look better and reduces tram noise, providing residents with a quieter life (Anon. (2005) Sustainability, Building, 18 November, pp. 74–75).
49 Man's inhumanity to man (woman and child) is epitomised by the type of housing that has not a single tree (as in the 'homes' for workers in the nineteenth century in some of the mill towns of Lancashire in the UK). In areas around the world at the moment other environments, where people will live without knowing pleasure, are being created.
50 Schiff, S. (2005) *Dr Franklin Goes to France*, London: Bloomsbury, p. 45.
51 The mass transport system may be supplied principally by renewable energy sources from outside the city. Already one sees signs of this in London where all the electricity for London Underground's offices and stations comes from renewable supplies (Anon. (2006) Tube Reduces Energy Use, Metro, 16 January, p. 52).
52 Anon. (2003) An Energy Revolution Solution for Sustainable Urban Communities. INREB Faraday Partnership International Design Ideas Competition, BRE, Garston. The team consisted of Richard Partington Architects, MacCormac Jamieson Pritchard, Luke Engleback and Max Fordham LLP.

Appendix A: Earth, sky and physics

1 Gribbin, J. (1988) 'The greenhouse effect', *New Scientist*, 22 October; *Inside Science*, no. 13.
2 Anon. (2005) 'Oceans are hiding climate time bomb', *New Scientist*, 7 May, p. 14. To put this in other contexts, the heat flux at the surface of the Earth derived from within the body of our planet is about 0.1 W/m^2 (44×10^{12} W divided by about 510×10^{12} m^2, from Fortey, R.

(2005) *The Earth An Intimate History*, London: Harper Perennial. Compare this with the heat loss from a person of about 55 W/m^2 in Figure 5.6.

3 Ibid.

4 Slezec, A.-M., Janvier, P. and Van Praet, M. (eds) (1991) *On a Marché Sur la Terre*, Paris: Editions ICS/Museum Nationale d'Histoire Naturelle, p. 216.

5 Another example of a large-scale effect is the melting of the polar ice caps. Polar ice reflects light and heat back into space. As more of it melts, more of the sun's energy is absorbed by the ocean with the potential effect of accelerating global warming.

6 Anon. (2006) 'Weather report', *New York Times*, 29 January, p. 33.
The brown clouds of smog (black carbon, organic carbon, and other aerosols such as sulphates and nitrates) formed by wildfires and burning fossil fuels and biofuels over Asia (Aldhaus, P. (2005). *Nature. China's Burning Ambition*, 435 (7046), pp. 1152–1154) are an example of a less beautiful sky. Black carbon, a by-product of coal-burning, absorbs solar radiation and in addition to reducing the light available results in a hotter atmosphere and cooler ground. These clouds are ill omens for all of us and it is important that worldwide environmental strategies be developed with India and China.

7 Anon. (2003) 'How much, how hot?' *Guardian*, 12 July, p. 23.

8 Anon. (2003) 'General situation', *The Times*, 17 October.

9 Bunn, R. (1998) 'Ground coupling explained', *Building Services*, December, pp. 22–27.

10 Long-tailed tits (one of the smallest of British birds) huddle together on long, cold winter nights and 'form a single feathered clump with numerous protruding tails'. More than 60 wrens have been observed in Norfolk crowded into a nest box measuring 14 × 11.5 × 14.5 cm. See Cocker, M. and Mabey, R. (2005) *Birds Brittanica*, London: Chatto and Windus, pp. 337 and 386.

11 CIBSE Guide A3 Thermal and other properties of building structures, 1977.

12 Anon. (1969) 'Condensation', *BRE Digest 110*, Garston: Building Research Station.

13 www.monachos.grden/resources/thermo/conductivity.asp.21/12/04.

14 Based on date in Monteith, J. (1973) *Principles of Environmental Physics*, London: Edward Arnold.

15 Thomas, R. (2006) *Environmental Design*, London: Spon Press.

16 Based on Schmitt, D. *et al.* (2003) 'Experimental evidence concerning spear use in Neanderthals and early modern humans', *Journal of Archaeological Science*, vol. 30, no. 1, pp. 103–114.

17 Ibid.

18 See reference in note 15, p. 11.

19 Moon, P. (1940) 'Proposed standard solar radiation curve for engineering use', *Journal of Franklin Institute*, November, p. 604.

20 Anon. (1994) *CIBSE Code for Interior Lighting*, London: CIBSE.

Appendix B: a time line

1 Stringer, C. (2003) *Nature*, 12 June, pp. 692–694.

2 August, O. and Hammond, N (2002) 'Chinese dig up relics from "majestic" town of 6000 BC', *The Times*, 3 October, p. 18.

3 Williams, Z. (2006) 'Call that risky?', *Guardian Weekend*, 1 April 2006, p. 34.

Bibliography

Bachelard, G., *The Poetics of Space*, Beacon Press, Boston, 1969.

Banham, R., *Theory and Design in the First Machine Age*, Architectural Press, London, 1960.

Banham, R., *The Architecture of the Well-tempered Environment*, Architectural Press, London, 1969.

Battle, G. and McCarthy, C., *Sustainable Ecosystems and the Built Environment*, Wiley-Academy, Chichester, 2001.

Blackburn, S., *The Oxford Dictionary of Philosophy*, Oxford University Press, Oxford, 1994.

Brace Taylor, B., *Le Corbusier: The City of Refuge Paris*, Chicago University Press, Chicago, 1984.

Brody, H., *The Other Side of Eden*, Faber & Faber, London, 2001.

Brownlee, D.B. and De Long, D.G., *Louis I. Kahn: In the Realm of Architecture*, Rizzoli, New York, 1993.

Campbell, J., 'Bios and Mythos', in *The Flight of the Wild Gander*, Harper Perennial, New York, 1990, pp. 27–42.

Craddock, N., *Light in the Work of Sir John Soane*, unpublished M.Phil., Cambridge University, 1995.

Darvill, D. and Thomas, J. (eds), *Neolithic Houses in North-west Europe and Beyond*, Oxbow Books, Oxford, 1996.

Dennis, M., *Court and Garden: From the French Hôtel to the City of Modern Architecture*, MIT Press, Cambridge, MA, 1998.

Drew, P., *Leaves of Iron*, The Law Book Co., Sydney, 1985.

Dunn, M., 'Japan', in G. Fahr-Becker (ed.), *The Art of Asia*, Köneman, Cologne, 1999.

Eliade, M., *The Sacred and the Profane*, Harcourt, Brace, Jovanovich, New York, 1959.

Eliade, M., *The Myth of the Eternal Return*, Princeton University Press, Princeton, 1971.

Elvin, M., *The Retreat of the Elephants: An Environmental History of China*, Yale University Press, New Haven, CT, 2004.

Ewart Evans, G., *Ask the Fellows who Cut the Hay*, Faber & Faber, London, 1956.

Farmer, J., 'Battered bunkers', *Architectural Review*, vol. 181 (January 1987), pp. 60–65.

Fathy, H., *Natural Energy and Vernacular Architecture*, University of Chicago Press, Chicago, 1986.

Frampton, K., *Studies in Tectonic Culture*, MIT Press, Cambridge, MA, 1995.
Frankl, G., *Archaeology of Mind*, Open Gate Press, London, 1992.
Garnham, T., *Melsetter House*, Phaidon, London, 1993.
Garnham, T., *Lines on the Landscape, Circles from the Sky*, Tempus, Stroud, 2004.
Giedion, S., *Mechanisation Takes Command*, Oxford University Press, Oxford, 1948.
Gould, S.J., *Ever Since Darwin: Reflections in Natural History*, Penguin, London, 1978.
Hamblyn, R., *The Invention of Clouds*, Picador, London, 2002.
Hawkes, D. and Forster, W., *Architecture, Engineering and Environment*, Laurence King Publishing, London, 2002.
Hawkes, J., *Man and the Sun*, The Cresset Press, London, 1962.
Hawkes, J., *Prehistory*, George Allen & Unwin, London, 1963.
Heaney, S., *Finders Keepers: Selected Prose 1971–2001*, Faber & Faber, London, 2002.
Heidegger, M., 'Building dwelling thinking', in *Poetry, Language, Thought*, Harper & Row, New York, 1975.
Heidegger, M., *The Question Concerning Technology*, Harper & Row, New York, 1977.
Hix, J., *The Glasshouse*, Phaidon, London, 1996.
Hodder, I., *The Domestication of Europe*, Blackwell, Oxford, 1990.
Hopkinson, R.G., Petherbridge, P. and Longmore, J., *Daylighting*, Heinemann, London, 1966.
Hoskins, W.G., *The Making of the English Landscape*, Penguin Books, Harmondsworth, 1971.
Imbert, B., *North Pole, South Pole: Journeys to the Ends of the Earth*, Thames and Hudson, London, 1992.
Le Corbusier, *Towards a New Architecture*, Architectural Press, London, 1946.
Le Corbusier, *Le Poème de l'Angle Droit*, Le Corbusier Foundation, Paris, 1989.
Le Corbusier, *Precisions*, MIT Press, Cambridge, MA, 1991.
Lethaby, W., *Philip Webb and his Work*, Raven Oak Press, London, 1979.
Liebel, B. and Brodwick, J., 'Choosing the right light', *ASHRAE*, vol. 47, no. 12, 2005, pp. 122–123.
Lloyd, N., *The History of the English House*, Architectural Press, London, 1976.
Lobell, J., *Between Silence and Light*, Shambala, Boulder, 1979.
Marzluff, J.M. and Angell, T., *In the Company of Crows and Ravens*, Yale University Press, New Haven, CT, 2005.
Mellart, J., *Çatal Hűyűk*, Thames and Hudson, London, 1967.
Messham, S., *Gas an Energy Industry*, HMSO, London, 1976.
Mithen, S., *The Prehistory of Mind*, Phoenix, London, 1998.
Morse, E., *Japanese Houses and their Surroundings*, Dover, NY, 1961.
Muir, E., *An Autobiography*, Canongate Press, Edinburgh, 1993.
Norberg-Shultz, C., *Genius Loci: Towards a Phenomenology of Architecture*, Academy Editions, London, 1980.
Oliver, P. (ed.), *The Encyclopaedia of Vernacular Architecture*, vol. 2, Cambridge University Press, Cambridge, 1997.
Oliver, P., *Dwellings: The Vernacular House Worldwide*, Phaidon, London, 2003.
Pascal, A. *et al.*, 'Molecular basis of photoprotection and control of photosynthetic light-harvesting', *Nature*, vol. 436 (July 2005), pp. 134–137.

Rapoport, A., *House Form and Culture*, Prentice Hall, New Jersey, 1969.

Redfield, R., *The Primitive World and its Transformation*, Penguin, London, 1968.

Renfrew, C. (ed.), *The Prehistory of Orkney*, Edinburgh University Press, Edinburgh, 1990.

Ruspoli, M., *The Caves of Lascaux*, Abrams, New York, 1987.

Rykwert, J., *On Adam's House in Paradise: the Idea of the Primitive Hut in Architectural History*, Museum of Modern Art, New York, 1972.

Schildt, G. (ed.), *Alvar Aalto in his own Words*, Rizzoli, New York, 1998.

Scully, V., *The Shingle and the Stick Style*, Yale University Press, New Haven, 1971.

Sennett, R., *The Conscience of the Eye: The Design and Social Life of Cities*, Faber & Faber, London, 1990.

Smith, P., *Architecture in a Climate of Change*, Butterworth-Heinemann, Oxford, 2001.

Snow, C.P., *The Masters*, Penguin, London, 1951.

van Eyck, A., 'The interior of time', in G. Baird and C. Jencks (eds), *Meaning in Architecture*, The Cresset Press, London, 1960.

Vitruvius, *The Ten Books on Architecture*, Dover, NY, 1960.

von Stimson, O., *The Gothic Cathedral*, The Bollingen Foundation, Pantheon Books, New York, 1962.

Wigglesworth, V.B., *The Life of Insects*, The New American Library, Weidenfeld & Nicolson, New York, 1966.

Wojciechowska, P., *Building with Earth*, Chelsea Green Publishing, White River Junction, VT, 2001.

Index

Note: page numbers in *italics* denote illustrations separated from the textual reference